Henning Meyer

Intersection Theory on Compact Tropical Toric Varieties

Henning Meyer

Intersection Theory on Compact Tropical Toric Varieties

and Compactifications of Tropical Parameter Spaces

Südwestdeutscher Verlag für Hochschulschriften

Imprint

Any brand names and product names mentioned in this book are subject to trademark, brand or patent protection and are trademarks or registered trademarks of their respective holders. The use of brand names, product names, common names, trade names, product descriptions etc. even without a particular marking in this work is in no way to be construed to mean that such names may be regarded as unrestricted in respect of trademark and brand protection legislation and could thus be used by anyone.

Cover image: www.ingimage.com

Publisher:
Südwestdeutscher Verlag für Hochschulschriften
is a trademark of
Dodo Books Indian Ocean Ltd., member of the OmniScriptum S.R.L Publishing group
str. A.Russo 15, of. 61, Chisinau-2068, Republic of Moldova Europe
Printed at: see last page
ISBN: 978-3-8381-2700-2

Zugl. / Approved by: Kaiserslautern, TU, Diss., 2011

Copyright © Henning Meyer
Copyright © 2011 Dodo Books Indian Ocean Ltd., member of the OmniScriptum S.R.L Publishing group

meiner Familie

Contents

Introduction		5
Chapter 1.	Toric Varieties	9
Chapter 2.	Tropical Intersection Theory	23
Chapter 3.	Tropicalization	43
Chapter 4.	Parameter Spaces of Lines in $\mathbb{TP}^n$	49
Chapter 5.	Chow Quotients	61
Chapter 6.	Rational Tropical Curves	69
Bibliography		91

Introduction

1. Introduction to Tropical Geometry

An affine algebraic variety is the zero set of finitely many polynomials. For example $X = \{(x_1, x_2) \in \mathbf{C}^2 \mid x_1^2 - x_2 = 0\}$ is a closed subset of real dimension two whose set of real points is the standard parabola. Tropical Geometry is concerned with the study of deformations of these varieties into polyhedral complexes:

If $X \subseteq \mathbf{C}^n$ is an algebraic variety, we can look at its amoeba

$$\mathcal{A}(X) = \{(\log_t|x_1|, \ldots, \log_t|x_n|) \mid x \in X, x_i \neq 0 \text{ for all } i\} \subseteq \mathbb{R}^n$$

for some $t > 0$.

The logarithmic limit set (or tropicalization) of X is the Hausdorff limit of these sets for $t \to 0$. It is a connected polyhedral complex of pure (real) dimension d when X is an irreducible variety of complex dimension d (see Figure 1(a)).

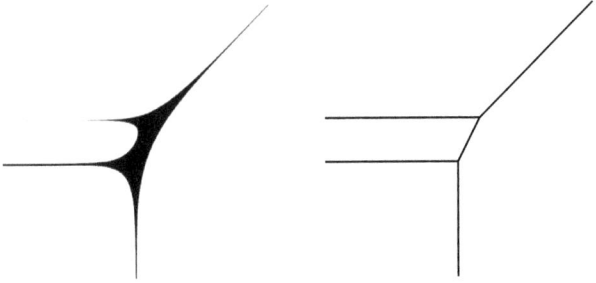

(a) The amoeba $\mathcal{A}(C)$ of the complex curve $C = \{(x,y) \in (\mathbf{C}^\times)^2 \mid x^2 + y^2 + 4x + 1 = 0\}$. For this image the base of the logarithm was chosen as $t = \sqrt{2}$.

(b) The non-Archimedean amoeba of the curve $C(\mathbb{K}) = \{(x,y) \in (\mathbb{K}^\times)^2 \mid x^2+y^2+ t^4 x + 1 = 0\}$ over the field $\mathbb{K} = \mathbf{C}\{\{t\}\}$ of complex Puiseux series.

FIGURE 1. A complex amoeba and a non-Archimedean amoeba

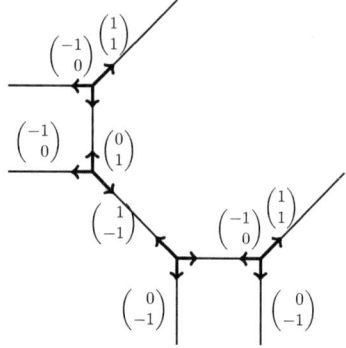

FIGURE 2. A tropical curve in $\mathbb{R}^2$. At every vertex the sum of the outgoing vectors is zero.

Instead of taking a limit of logarithms of the usual Euclidean absolute value, the modern approach studies the set

$$\mathcal{A}(X(\mathbb{K})) = \{(\operatorname{val} x_1, \ldots, \operatorname{val} x_n) \mid x \in X(\mathbb{K}), x_i \neq 0 \text{ for all } i\}$$

where $\mathbb{K}$ is an algebraically closed field extending $\mathbb{C}$ with a non-Archimedean valuation val, i.e. a group homomorphism val : $\mathbb{K}^\times \to \mathbb{R}$ that satisfies the ultra-metric triangle inequality

$$\operatorname{val}(a+b) \leq \max(\operatorname{val}(a), \operatorname{val}(b)).$$

The set $X(\mathbb{K})$ is defined as all points of $\mathbb{K}^n$ that satisfy the same equations as X. In this case the set $\mathcal{A}(X(\mathbb{K}))$ is a polyhedral complex and called the non-Archimedean amoeba of X (see Figure 1(b) on the previous page).

A crucial feature of these polyhedral complexes is that they satisfy a balancing condition (sometimes called a zero-tension condition) at every cell of codimension one (see Figure 2).

2. Overview of Thesis and Main Results

This work can be subdivided into two parts:

- The first part develops an intersection theory for tropical cycles in toric varieties. This part contains chapters one up to three. The main results are in Sections 2.3 and 2.4, while the rest of chapter 2 is devoted to presenting the already existing theory.
- The second part describes the combinatorics of certain toric compactifications of parameter spaces for tropical curves. It consists of chapters four to six. The main results are in chapter four and chapter six. Chapter 4 investigates the tropical Grassmannian, with emphasis on the Grassmannian of lines. Chapter 5 collects results about

Chow quotients and fiber polytopes. These are used in Chapter 6 to construct compactifications of the tropical parameter spaces of n-marked rational curves of degree d.
- Chapters one and three, which develop tropical toric varieties and the relation to toric varieties over non-Archimedean fields are relevant for both parts and might be of independent interest.

In Chapter 1 we construct tropical toric varieties in complete analogy to the complex case (for which [**Ful93**] is the standard reference). If $\mathbb{K}$ is an algebraically closed field with a non-Archimedean valuation, then we can consider a tropicalization map from a toric variety over $\mathbb{K}$ to the corresponding tropical toric variety, extending the usual tropicalization from the algebraic torus $(\mathbb{K}^\times)^n$ to $\mathbb{R}^n$ (as in [**Pay09a**]).

In Chapter 2 we develop a theory of tropical cycles inside a tropical toric variety, generalizing the theory of tropical cycles inside $\mathbb{R}^n$ as described in [**AR09**].

For complete smooth toric varieties we are able to construct an intersection theory of these cycles that unifies the stable intersection of tropical varieties, the intersection of Minkowski weights and the intersection of torus invariant subvarieties.

We then focus on compactifications of tropical fans inside tropical toric varieties.

We study the combinatorics of these compactifications for several spaces related to tropical Grassmannians (Chapter 4): The parameter spaces $\mathrm{M}_{0,n}^{\mathrm{lab}}(\mathbb{R}^r, d)$ of labeled n-marked tropical rational curves of degree d inside $\mathbb{R}^r$ from [**GKM09**] (they are quotients of the tropical Grassmannian).

In Chapter 6 we describe a compactification $\overline{\mathrm{M}}_{0,n}^{\mathrm{lab}}(\mathbb{TP}^r, d)$ whose boundary points correspond to connected tropical curves of genus zero and degree d with n marked points in $\mathbb{TP}^r$. We construct this compactification by taking a Chow quotient of the rank two tropical Grassmannian by a linear subspace of its lineality space.

We use methods similar to those of [**Kap93**] and [**GM07**] to study the combinatorics of the corresponding Chow quotients of complex varieties.

Acknowledgements. I would like to thank Andreas Gathmann, Bernd Sturmfels, Carolin Torchiani, Christian Haase, Dennis Ochse, George François, Hannah Markwig, Johannes Rau, Kristin Shaw, Lars Allermann, Maike Lorenz, Sarah Brodsky and Simon Hampe. My stay at the Tropical Geometry program of the Mathematical Sciences Research Institute has been very helpful for furthering this thesis.

CHAPTER 1

Toric Varieties

In this section we will construct tropical toric varieties, tropical analogues to complex toric varieties. We begin with a short review of complex toric varieties. Those are algebraic varieties constructed from polyhedral data in such a way that the resulting variety has combinatorics similar to the polyhedral data.

Definition 1.1. Let $N \cong \mathbb{Z}^n$ be a lattice and $V = N \otimes \mathbb{R}$ the corresponding real vector space. The intersection of finitely many halfspaces in V is called a polyhedron. Such a polyhedron is usually written as $P = \{x \in V \mid Ax \geq b\}$ where A is a vector in $(V^\vee)^r$ and b in $\mathbb{R}^r$ (usually we have $V \cong \mathbb{R}^n$, then A is an $r \times n$-matrix).

If A lies in the lattice $(N^\vee)^r$ and b in the lattice $\mathbb{Z}^r$ then P is called a rational polyhedron. If A lies in $(N^\vee)^r$ but b is only in $\mathbb{R}^r$ then P is called a polyhedron with rational slopes.

If P is a polyhedron and $f \in V^\vee$ a linear form and $a \in \mathbb{R}$ with $f \cdot p \geq a$ for all $p \in P$ then the set
$$\{p \in P \mid f \cdot p = a\}$$
is called a face of P.

If τ is a non-empty polyhedron, we use the notation $\sigma > \tau$ to denote that τ is a face of σ and $\dim(\tau) + 1 = \dim(\sigma)$. We call τ a facet of σ in this case.

Theorem 1.2. Let M be any finite point set in a real vector space. Then
$$\operatorname{conv}(M) := \left\{\sum \lambda_i m_i \mid m_i \in M, \lambda_i \in [0,1], \sum \lambda_i = 1\right\}$$
and
$$\operatorname{pos}(M) := \left\{\sum \lambda_i m_i \mid m_i \in M, \lambda_i \geq 0\right\}$$
are polyhedra. Every polyhedron is of the form $\operatorname{conv}(A) + \operatorname{pos}(B)$ for some finite sets A, B.

PROOF. This is a standard result about polyhedra, see for example chapter one of [**Zie95**]. □

The objects that we need are polyhedral fans, collections of polyhedra satisfying certain compatibility conditions.

Definition 1.3. A polyhedron C is a polyhedral cone if $\lambda x \in C$ for all $\lambda > 0$ and $x \in C$. In other words, C is a cone if $C = \text{pos}(C)$. It is a pointed cone if the origin is a face. A compact polyhedron is called a polytope.

A polyhedral complex G is a set of polyhedra such that for all U in G all faces of U are in G and for all U, V in G the intersection $U \cap V$ is a face of both U and V.

We use the notation $F^{(k)}$ to denote the set of k-dimensional polyhedra of a polyhedral complex F and $|F|$ to denote the underlying set $|F| := \bigcup_{\sigma \in F} \sigma$.

A polyhedral fan is a polyhedral complex such that all polyhedra are cones. A rational fan is a polyhedral fan such that all cones are rational polyhedra. A polyhedral fan F in a vector space V is complete if $|F| = V$.

The bridge from polyhedral geometry to algebra will be via semigroups derived from pointed rational polyhedral cones.

Definition 1.4 (Semigroup, Semifield).

(1) A semigroup is a set S together with an associative binary operation
$$\cdot : S \times S \to S.$$

(2) A map $f : S \to T$ between semigroups is a morphism of semigroups if $f(ab) = f(a)f(b)$ for all a, b in S.

(3) A semigroup $(S, \cdot)$ is commutative if the operation $\cdot$ is commutative.

(4) An element e of a commutative semigroup $(S, \cdot)$ is called a neutral element if $e \cdot s = s \cdot e = s$ for all $s \in S$.

(5) A semigroup is a cancellative semigroup or a monoid if it is isomorphic as semigroup to a subset of a group G.

(6) A semigroup $(S, \cdot)$ is called idempotent if $a \cdot b \in \{a, b\}$ for all $a, b \in S$. A monoid cannot be idempotent (unless it consists only of the neutral element). The addition of the tropical semifield defined below is such an idempotent operation.

(7) A set S with two associative operations $\oplus : S \times S \to S$ and $\odot : S \times S \to S$ is called a semiring if $(S, \oplus)$ is a commutative semigroup with neutral element and $\odot$ is distributive over $\oplus$.

(8) Let $(S, \oplus, \odot)$ be a semiring with neutral element e (respective to $\oplus$). Then S is a semifield if $(S \setminus \{e\}, \odot)$ is an Abelian group.

All semigroups in this work will be commutative with a neutral element.

Definition 1.5 (Tropical Semifield $\mathbb{T}$). The set $\mathbb{T} = \mathbb{R} \cup \{-\infty\}$ is the semifield of tropical numbers with operations
$$\oplus : \mathbb{T} \times \mathbb{T} \to \mathbb{T}, (a, b) \mapsto a \oplus b = \max(a, b)$$

and
$$\odot : \mathbb{T} \times \mathbb{T} \to \mathbb{T}, (a,b) \mapsto a \odot b = a + b.$$
As a topological space, $\mathbb{T}$ carries the topology of the half-open interval $[0,1[\approx [-\infty, +\infty[$.

Example 1.6.

- Every group is a semigroup, every ring a semiring and every field a semifield.
- Let $\mathbb{K} = (\mathbb{K}, +, \cdot)$ be a field. We write $\mathbb{K}^\times$ for the multiplicative group $(\mathbb{K} \setminus \{0\}, \cdot)$ of $\mathbb{K}$. Then $(\mathbb{K}, \cdot) = (\mathbb{K}^\times \cup \{0\})$ is a semigroup that is not a monoid.
- Let $C \subseteq \mathbb{R}^n$ be a polyhedral cone. Then the sets $C, C \cap \mathbb{Q}^n$ and $C \cap \mathbb{Z}^n$ are monoids.
- The absolute value $|\cdot| : (\mathbb{C}, \cdot) \to (\mathbb{R}_{\geq 0}, \cdot)$ is a morphism of semigroups. If $\mathbb{K}$ is a non-Archimedean field, then the valuation $\mathrm{val} : \mathbb{K} \to \mathbb{T}$ is a morphism of semigroups.

Definition 1.7. Let $\mathbb{K}$ be a semifield. The n-dimensional (algebraic) torus over $\mathbb{K}$ is the set $(\mathbb{K}^\times)^n$.

Definition 1.8. A toric variety is a pair (T, X) where X is an irreducible algebraic variety over a field $\mathbb{K}$ and T is an algebraic torus acting on X such that there exists an open T-orbit of X isomorphic to T.

The torus T comes with a lattice of one-parameter sub-groups
$$N = \hom(\mathbb{K}^\times, T) := \{\lambda : \mathbb{K}^\times \to T \mid \lambda \text{ is a continuous group homomorphism}\}$$
and a dual lattice of characters
$$N^\vee = \hom(T, \mathbb{K}^\times) := \{\chi : T \to \mathbb{K}^\times \mid \chi \text{ is a continuous group homomorphism}\}.$$

If our torus is $T = (\mathbb{K}^\times)^n$ then the lattice N is given by $\hom((\mathbb{K}^\times)^n, \mathbb{K}^\times) = \mathbb{Z}^n$.

Let us assume that we have $\mathbb{K} = \mathbb{C}$ and let us assume that (T, X) is a complex toric variety that is compact in the Euclidean topology.

That means for every one-parameter subgroup λ the limit $\lim_{t \to 0} \lambda(t)$ exists. We can define an equivalence relation on the lattice $N^\vee$ such that two one-parameter subgroups are equivalent if they have the same limit point. It turns out that the corresponding equivalence classes are (lattice points of relative interiors of) polyhedral cones. Thus we get a fan structure on N or rather the vector space $N \otimes \mathbb{R}$ (an example of this is worked out in [**Cox01**]). This fan structure determines the topology of the toric variety (T, X).

It is a result of [**Oda78**, Theorem 4.1] that a normal toric variety is determined uniquely by its fan, and we will now focus on the synthetic construction of a toric variety from a fan.

Will now describe a construction of toric varieties over arbitrary semifields. In most parts, this is completely analogous to the theory of toric varieties over $\mathbb{C}$ as described in [**Ful93, Ewa96**].

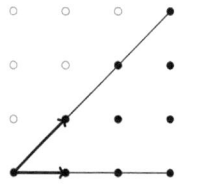 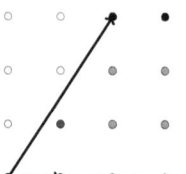

(a) Every lattice point in a unimodular cone is a positive integer combination of the integer vectors generating the cone.

(b) A cone that is not unimodular. The semigroup needs an interior point as additional generator.

FIGURE 3. The semigroups of a unimodular cone and a cone that is not unimodular.

However, there is no Spec and no commutative rings, so the constructions will occasionally be less elegant than in the classical theory.

The relationship between toric varieties and tropical geometry has been known before. Many authors relate the vector space $\mathbb{R}^n$ to the torus $(\mathbb{K}^\times)^n$ of a non-Archimedean field, but few consider the extension $\mathbb{R}^n \subseteq \mathbb{T}^n$ since the infinite points of $\mathbb{T}^n$ interfere with the polyhedral geometry in $\mathbb{R}^n$. Nonetheless, toric varieties over the tropical semifield have been considered before, most notably by Sam Payne in [**Pay09a**] and by Takeshi Kajiwara (unpublished, but announced in [**HKMP06**]).

We will connect the intersection theory of [**AR09**] inside the torus $\mathbb{R}^n$ with the usual intersection theory of torus-invariant subspaces and the description of cohomology classes via Minkowski weights from [**FS97**].

Definition 1.9. Let $\sigma \subseteq V$ be a pointed rational cone and
$$\sigma^\vee := \{x \in V^\vee \mid x \cdot s \geq 0 \text{ for all } s \in \sigma\}$$
the dual cone. Then $S_\sigma := \sigma^\vee \cap M$ is a finitely generated semigroup by Gordan's Lemma (see e.g. [**Ful93**, Prop 1.1]). If τ is a face of σ, then the inclusion of sets $i : S_\sigma \to S_\tau$ is a morphism of semigroups.

We call the cone σ unimodular if it is generated by a subset of a basis of N and simplicial if it can be generated by a linearly independent subset of N.

Remark 1.10. We require σ to be pointed so that $\sigma^\vee$ is full-dimensional.

There are two basic constructions of toric varieties in complex algebraic geometry:

- To a rational fan F one can associate a normal toric variety $\mathbf{X}_F(\mathbf{C})$ which is covered by affine sets depending on the cones of the fan F.

- To a collection $\mathcal{A} = \{a_1, \ldots, a_k\} \subseteq \mathbb{Z}^d$ of lattice points with d-dimensional convex hull we can associate a projective toric variety $\mathbf{Y}_\mathcal{A}(\mathbb{C})$. The matrix $A = (a_1 \ldots a_k)$ defines a map $(\mathbb{C}^\times)^d \to (\mathbb{C}^\times)^k$ via $z \mapsto (z^{a_j})_j$. Now $\mathbf{Y}_\mathcal{A}(\mathbb{C})$ is the closure of the image of $(\mathbb{C}^\times)^d$ in $\mathbb{P}(\mathbb{C}^k)$.

If the set $\mathcal{A}$ is the set of vertices of a full-dimensional polytope P with normal fan F, then $\mathbf{Y}_\mathcal{A}(\mathbb{C}) = \mathbf{X}_F(\mathbb{C})$.

We will start with a tropical analogue $\mathbf{X}_F(\mathbb{T})$ of $\mathbf{X}_F(\mathbb{C})$. The basic building blocks will be the set $\hom(S_\sigma, \mathbb{T})$. We will equip $\hom(S_\sigma, \mathbb{T})$ with a topology (as a subspace of $\mathbb{R}^D$ for some D). Furthermore, it contains the vector space $\hom(S_\sigma, \mathbb{R})$ as a dense open subset and contains the lattice $\hom(S_\sigma, \mathbb{Z})$.

Definition 1.11. For $x \in \mathbb{R}$ let $x^+ := \max(x, 0)$ and $x^- := -\min(0, x)$, i.e. $x = x^+ - x^-$ and both numbers are non-negative. We use the same notation component-wise for matrices and vectors.

We will use this when considering linear equations over $\mathbb{T}$:

Let f be a vector in $\mathbb{R}^{1 \times n}$ and $x \in \mathbb{R}^n$. The equations $f \cdot x = 0$ and $f^+ \cdot x = f^- \cdot x$ are equivalent, but with $x \in \mathbb{T}^n$ only the latter expression is defined for all x.

Definition 1.12. Let $A \subseteq \mathbb{Z}^d$ be a semigroup with a finite set of generators $G = \{g_1, \ldots, g_k\}$. Let $R = \{r_1, \ldots, r_n\} \subseteq \mathbb{Z}^k$ generate the integer relations between the g_i, i.e. $\mathrm{span}_\mathbb{Z}(R) = \{z \in \mathbb{Z}^k \mid \sum g_i z_i = 0\}$. Let D be another commutative semigroup. We define

$$K(G, R, D) := \left\{ x \in D^{|G|} \mid r^+ \cdot x = r^- \cdot x \ \forall r \in R \right\}.$$

If D is a group then $K(G, R, D)$ is a group and if it is a ring then $K(G, R, D)$ is a D-module.

Lemma 1.13.

(1) Let $A \subseteq \mathbb{Z}^d$ be a finitely generated semigroup and $G = \{g_1, \ldots, g_k\}$ a set of generators with relations generated by $R = \{r_1, \ldots, r_n\}$. Let D be an additive semigroup. Then $\hom(A, D)$ is in bijection with $K(G, R, D)$.

(2) Let $D = \mathbb{T}$. If $H = \{h_1, \ldots, h_l\}$ is another set of generators with relations $S = \{s_1, \ldots, s_m\}$ then there is a linear isomorphism $K(G, R, \mathbb{R}) \to K(H, S, \mathbb{R})$ that extends to a homeomorphism $K(G, R, \mathbb{T}) \to K(H, S, \mathbb{T})$ and restricts to a group isomorphism $K(G, R, \mathbb{Z}) \to K(H, S, \mathbb{Z})$.

PROOF. We first show the inclusion $\hom(A, D) \subseteq K(G, R, D)$. Let f be an element in $\hom(A, D)$. Define $x \in D^G$ via $x_i = f(g_i)$. We have $r^+ g_i = r^- g_i$ for all $g_i \in G$ and $r \in R$, hence $r^+ f(g_i) = f(r^+ g_i) = f(r^- g_i) = r^- f(g_i)$, that means

$$x \in \left\{ y \in D^G \mid r^+ \cdot y = r^- \cdot y \ \forall r \in R \right\}.$$

Now we show the other inclusion. Let $x \in K(G, R, D)$. Let $s \in A$. There is a representation $s = \sum a_i g_i$ with $a_i > 0$. Define $f : A \to \mathbb{T}$ via $f(s) = \sum a_i x_i$. We need to show that f is well-defined:

Assume $s = \sum b_i g_i$ is another representation. We know $\sum a_i g_i = \sum b_i g_i$ and we want to show $\sum a_i x_i = \sum b_i x_i$. Now, since a and b are vectors in $\mathbb{Z}^G$ we can look at their difference $a - b$. We have $\sum (a_i - b_i) g_i = 0$ since $\sum a_i g_i = \sum b_i g_i$. But that means $(a_i - b_i)_i$ is an integer relation on the g_i, hence $\sum (a_i - b_i)^+ x_i = \sum (a_i - b_i)^- x_i$ which means $\sum a_i x_i = \sum b_i x_i$.

Now we want to prove the second statement of the lemma. Let g_i and h_j be two different generating systems of A and let the base change be given via the relations $g_i = \sum \lambda_{ij} h_j$ and $h_j = \sum \mu_{ji} g_i$ with $\lambda_{ij}, \mu_{ji} \in \mathbb{Z}_{\geq 0}$. We want to show: the map $T : K(G, R, \mathbb{R}) \to K(H, S, \mathbb{R}), x \mapsto (\sum_i \mu_{ji} x_i)_j$ is a linear homeomorphism with inverse $T^{-1} : K(H, S, \mathbb{R}) \to K(G, R, \mathbb{R}), y \mapsto (\sum_j \lambda_{ij} y_j)_i$. We know $g_i = \sum \lambda_{ij} \sum \mu_{jk} g_k$ and $h_j = \sum \mu_{ji} \sum \lambda_{ik} h_k$, therefore
$$1 \cdot h_j = \sum_k (\sum_i \mu_{ji}) \lambda_{ik} h_k$$
is an integer relation on the h_j. Hence $y_j = \sum_k (\sum_i \mu_{ji}) \lambda_{ik} y_k$ for all $y \in K(H, S, \mathbb{R})$ and similarly $x_i = \sum_k (\sum_j \lambda_{ij}) \mu_{jk} x_k$ for all $x \in K(G, R, \mathbb{R})$.

So $x \stackrel{T}{\mapsto} (\sum_i \mu_{ji} x_i)_j \stackrel{T^{-1}}{\mapsto} (\sum_j \lambda_{ij} \sum_i \mu_{ji} x_i)_i$ which means the maps are inverse to each other (the other direction follows from symmetry). By the same reasoning, we get a group isomorphism $K(G, R, \mathbb{Z}) \to K(H, S, \mathbb{Z})$ and a bijection $K(G, R, \mathbb{T}) \to K(H, S, \mathbb{T})$. □

Definition 1.14 (Affine toric variety U_σ). Let σ be a cone and S_σ the corresponding semigroup. We define $U_\sigma := U_\sigma(\mathbb{T}) := \hom(S_\sigma, \mathbb{T})$. We equip U_σ with the subspace topology induced via an embedding as in the preceding lemma.

Remark 1.15. Let A be a subsemigroup of $\mathbb{Z}^n$. Then the set $\hom(A, \mathbb{T})$ contains the real vector space $\hom(A, \mathbb{R})$ and the lattice $\hom(A, \mathbb{Z}) \cong \mathrm{span}_{\mathbb{Z}}(A)$.

Remark 1.16. If $\mathbb{K}$ is a field then $U_\sigma(\mathbb{K}) = \hom(S_\sigma, \mathbb{K})$ is isomorphic to the closed points of the scheme $\mathrm{Spec}\,\mathbb{K}[S_\sigma]$. Toric varieties over $\mathbb{C}$ are usually considered as analytic spaces with the Euclidean topology. If $\mathbb{K}$ is a non-Archimedean field then $\mathbb{K}$ has a topology induced by the valuation. This topology, however, turns $\mathbb{K}$ into a totally disconnected topological space. Usual remedies are the use of Grothendieck topologies in rigid analytic geometry or the embedding of $\mathbb{K}$ and varieties over $\mathbb{K}$ into the corresponding Berkovich spaces [**Ber90**].

Remark 1.17. We can describe the topology of $U_\sigma(\mathbb{T})$ in terms of σ: $U_\sigma(\mathbb{T})$ is homeomorphic to $\sigma^\vee$ as a cell complex. We will prove this in Theorem 1.33.

Example 1.18. Let $\sigma = \text{pos}\left\{\begin{pmatrix}-1\\2\end{pmatrix}, \begin{pmatrix}2\\-1\end{pmatrix}\right\}$. The dual cone is given via $\sigma^\vee = \text{pos}\left\{\begin{pmatrix}1\\2\end{pmatrix}, \begin{pmatrix}2\\1\end{pmatrix}\right\}$. These cones are simplicial but not unimodular.

Let us compare the lattice $\hom(\sigma^\vee \cap \mathbb{Z}^2, \mathbb{Z})$ with $\hom(\{0\}^\vee \cap \mathbb{Z}^2, \mathbb{Z}) = \mathbb{Z}^2$. A generating set for $\sigma^\vee \cap \mathbb{Z}^2$ is given by the vectors $v_1 = \begin{pmatrix}1\\2\end{pmatrix}$, $v_2 = \begin{pmatrix}2\\1\end{pmatrix}$, $v_3 = \begin{pmatrix}1\\1\end{pmatrix}$. They satisfy the relation $v_1 + v_2 = 3v_3$. Any tuple $(z_1, z_2, z_3) \in \mathbb{Z}^3$ satisfying this relation gives rise to a map $f : \sigma^\vee \cap \mathbb{Z}^2 \to \mathbb{Z}$ via $p = \alpha_1 v_1 + \alpha_2 v_2 + \alpha_3 v_3 \mapsto \alpha_1 z_1 + \alpha_2 z_2 + \alpha_3$.

The inclusion $\mathbb{Z}^2 = \hom(\mathbb{Z}^2, \mathbb{Z}) \to \hom(\sigma^\vee \cap \mathbb{Z}^2, \mathbb{Z})$ maps (x, y) to the tuple $(2x + y, 2y + x, x + y)$.

Remark 1.19. Assume that σ is unimodular. Let $b_1, \ldots, b_n$ be a basis of N with $\sigma = \text{pos}(b_1, \ldots, b_k)$. Then $\sigma^\vee = \text{pos}(b_1^\vee, \ldots, b_k^\vee, b_{k+1}^\vee, \ldots, b_n^\vee, -b_{k+1}^\vee, \ldots, -b_n^\vee)$. The only relation on these generators are generated by $b_{k+1}^\vee + (-b_{k+1}^\vee) = 0$. Hence

$$U_\sigma = \left\{ x \in \mathbb{T}^{n+n-k} \mid x_{k+i} + x_{n+i} = 0 \text{ for all } i = 1, \ldots, n-k \right\}.$$

This means x_{k+i} and x_{n+i} cannot be infinite for $i > k$. All coordinates x_{n+i} are determined via $x_{n+i} = -x_{k+i}$. All coordinates x_i with $i \leq k$ have no condition on them. Hence we see $U_\sigma \cong \mathbb{T}^k \times \mathbb{R}^{n-k}$.

Note that as $\sigma^\vee = \text{pos}(b_1^\vee, \ldots, b_k^\vee) + \text{span}(b_{k+1}^\vee, \ldots, b_n^\vee)$, we have $\sigma^\vee \cong \mathbb{R}_{\geq 0}^k \times \mathbb{R}^{n-k} \approx \mathbb{T}^k \times \mathbb{R}^{n-k} \cong U_\sigma$.

Definition 1.20 (Tropical Toric Variety). Let F be a rational fan and σ, τ cones of F. The inclusion $i_{\sigma,\tau} : S_\sigma \to S_\tau$ induces an inclusion $U_\tau \to U_\sigma$. We identify U_τ with $i_{\sigma,\tau}(U_\tau) \subseteq U_\sigma$ for all $\tau \subseteq \sigma$ and define the tropical toric variety as the topological space

$$\mathbf{X}_F(\mathbb{T}) := \coprod_{\sigma \in F} U_\sigma / \sim$$

where we glue along all those identifications $i_{\sigma,\tau}$.

We should view a tropical toric variety X as a triple (N, T, X) satisfying the following properties

- T is a dense open subset of the topological space X homeomorphic to a finite-dimensional real vector space.
- $N \subseteq T$ is a lattice and T is isomorphic to $N \otimes \mathbb{R}$ and $\hom(N, \mathbb{T})$.
- X has a finite open cover U_j. Each U_j contains T and is homeomorphic to a subset of some $\mathbb{T}^{n_j}$. In the case that F is complete and unimodular there is an open cover U_j such that each U_j is homeomorphic to $\mathbb{T}^n$ where $n = \dim T$.
- Every transition map preserves the vector space T and the lattice N. In the unimodular case a transition map is given by an invertible integer matrix.

- T acts on X extending the action on itself. We will later see that each orbit of the action is isomorphic to a quotient of T by a subspace and contains a lattice that is isomorphic to a quotient of N by the same subspace.
- A tropical toric variety X constructed from a rational fan F contains this fan in its torus T. Torus orbits will be in one-to-one correspondence with cones of F as in the complex case. We will later see that the closure of F in X is compact (even when X is not compact).

Example 1.21. There are (up to isomorphism, defined below) three fans in $V = \mathbb{R} = \mathbb{R}^1$.

$F_0 = \{0\}$ has only one cone, it corresponds to the toric variety $\mathbf{X}_{F_0}(\mathbb{T}) = U_{\{0\}} = \hom(\mathbb{Z}, \mathbb{T}) = \mathbb{R}$.

$F_1 = \{\{0\}, \mathbb{R}_{\geq 0}\}$ has one chain of cones $\mathbb{R} = U_{\{0\}} \subseteq U_{\mathbb{R}_{\geq 0}} = \hom(\mathbb{N}, \mathbb{T}) = \mathbb{T}$. Therefore $\mathbf{X}_{F_1}(\mathbb{T}) = \mathbb{T}$.

$F_2 = \{\mathbb{R}_{\leq 0}, \{0\}, \mathbb{R}_{\geq 0}\}$ has two maximal cones each isomorphic to $\mathbb{T}$. Their intersection is the torus $U_{\mathbb{R}_{\geq 0}} \cap U_{\mathbb{R}_{\leq 0}} = U_{\{0\}} \cong \mathbb{R}$. One embedding is given via $U_{\{0\}} \to U_{\mathbb{R}_{\geq 0}}, x \mapsto x$ as it comes from the inclusion $\mathbb{N} \to \mathbb{Z}$. The other embedding is given via $U_{\{0\}} \to U_{\mathbb{R}_{\leq 0}}, x \mapsto -x$ as it comes from the inclusion $-\mathbb{N} \to \mathbb{Z}$. We glue together two copies of $\mathbb{R} \cup \{-\infty\}$ over $\mathbb{R}$ via the identification $x \mapsto -x$. Hence $\mathbf{X}_{F_2}(\mathbb{T}) = \{-\infty\} \cup \mathbb{R} \cup \{+\infty\}$.

Definition 1.22. [Ewa96]

Let $\mathbf{X}_F$ be a toric variety over a semifield. A point $p \in \mathbf{X}_F$ is called regular or smooth if there is a unimodular cone $\sigma \in F$ such that $p \in U_\sigma$.

The toric variety $\mathbf{X}_F$ is called regular or smooth if every point is regular and singular otherwise.

Remark 1.23. $\mathbf{X}_F$ is regular as defined above if and only if F is unimodular. $\mathbf{X}_F(\mathbb{K})$ is regular as defined above if and only if the corresponding algebraic variety over $\mathbb{K}$ is regular. $\mathbf{X}_F(\mathbb{C})$ is regular as defined above if and only if it is smooth as a complex analytic space.

Definition 1.24 (Subfan)**.** A fan G is a subfan of a fan F if every cone of G is a cone of F.

Definition 1.25 (Map of fans)**.** Let $F \subseteq V \cong \mathbb{R}^n$ be a fan with lattice N, $G \subseteq V' \cong \mathbb{R}^m$ be another fan with lattice N'. Let $h : V \to V'$ be a linear map such that $h(N) \subseteq N'$ and $h(\sigma)$ is contained in a cone of G for every cone σ of F. We call h a map of fans and write $h : F \to G$.

Proposition 1.26. *Let $h : F \to G$ be a map of fans. Then h extends to a continuous map $h : \mathbf{X}_F(\mathbb{T}) \to \mathbf{X}_G(\mathbb{T})$.*

PROOF. Let $x \in U_\sigma$ and $\sigma' \in G$ with $h(\sigma) \subseteq \sigma'$. An element $m \in S_{\sigma'}$ determines a map $m : N' \to \mathbb{Z}, n \mapsto m \cdot n$.

Now it also determines a map $h^*(m) : N \to \mathbb{Z}, n \mapsto m \cdot h(n)$. If $n \in \sigma$ then $h(n) \in \sigma'$, hence

$m \cdot h(n) = 0$ and $h^*(m) \in S_\sigma$. Hence we have a map $h^* : S_{\sigma'} \to S_\sigma$ that induces a map $h_* : U_\sigma \to U_{\sigma'}$. These maps glue together to form a map $h : \mathbf{X}_F(\mathbb{T}) \to \mathbf{X}_G(\mathbb{T})$. □

Let $F \subseteq V$ be a rational fan with lattice N. As V is the torus of the toric variety $\mathbf{X}_F(\mathbb{T})$, there should be an action of V on $\mathbf{X}_F(\mathbb{T})$.

Definition 1.27. The action of $V = U_{\{0\}} = \hom(N^\vee, \mathbb{T})$ on $\hom(N^\vee \cap \sigma^\vee, \mathbb{T})$ is the addition of functions, the semigroup operation in $\hom(N^\vee \cap \sigma^\vee, \mathbb{T}) \supseteq \hom(N^\vee, \mathbb{T})$. It is the usual component-wise addition of points in $\mathbb{T}^n$ after the choice of a generating system of σ (as in Lemma 1.13). This defines an action of $U_{\{0\}}$ on all of $\mathbf{X}_F(\mathbb{T})$.

Theorem 1.28 (Torus Orbits). *Let $F \subseteq V$ be a rational fan with lattice $N \subseteq V$. There are decompositions*

(1)
$$U_\sigma = \coprod_{\tau \subseteq \sigma} O(\tau)$$

(2)
$$\mathbf{X}_F(\mathbb{T}) = \coprod_{\tau \in F} O(\tau)$$

into orbits $O(\tau)$ of the action of V on $\mathbf{X}_F(\mathbb{T})$ where $O(\tau) \subseteq \hom(\tau^\vee \cap N^\vee, \mathbb{T})$ is isomorphic to $N \otimes \mathbb{R}/\operatorname{span}(\tau)$.

Here $[v] \in N(\tau)$ corresponds to $\psi_{[v]} : \tau^\vee \cap N^\vee \to \mathbb{T}$ with

$$\psi_{[v]}(u) = \begin{cases} uv & \text{if } u \in \tau^\perp \\ -\infty & \text{otherwise.} \end{cases}$$

PROOF. Let f be an element of $\hom(N^\vee, \mathbb{T})$. If $[v] \in N \otimes \mathbb{R}/\operatorname{span}(\tau)$, then

$$(f + \psi_{[v]})(u) = \begin{cases} uv + f(u) & \text{if } u \in \tau^\perp \\ -\infty + f(u) & \text{otherwise.} \end{cases}$$

Hence $f\psi_{[v]} = \psi_{[f+v]} \in O(\tau)$, which means the $O(\tau)$ are orbits of the V-action.

Let $f \in \hom(N^\vee \cap \sigma^\vee, \mathbb{T})$. We need to find $\tau \subseteq \sigma$ and $[v] \in N \otimes \mathbb{R}/\operatorname{span}(\tau)$ such that $f = \psi_{[v]}$.

Let $G = \{g_1, \ldots, g_k\}$ be a generating system of σ.

If $W = f^{-1}(-\infty)$ is non-empty, then it is a sub-semigroup of S_σ. It must be generated by a subset H of G. The cone of H must be a face of σ (as every convex combination of points in W has image $-\infty$ under f). Hence we have a face $\tau \subseteq \sigma$ such that $f(u) = -\infty$ if and only if $u \in \operatorname{span}(\tau)$. Now f is a map to $\mathbb{R}$ on $M \cap \tau^\perp$ so it is defined by an element $[v] \in N/\operatorname{span}(\tau)$. Hence every element of $\hom(S_\sigma, \mathbb{T})$ corresponds to exactly one element in one $O(\tau)$, which means the $O(\tau)$ are precisely the torus orbits. □

Remark 1.29. The same result is true in the complex case [**Ful93**, Prop. 3.1] and could therefore be obtained via tropicalization. When tropical toric varieties were introduced in [**Pay09a**], they were defined as the disjoint union $\coprod_{\tau \in F} N(\tau)$ and then equipped with a global topology.

Theorem 1.30 (Orbit Closures). *Let F be a smooth complete fan. Let $O(\sigma)$ be a torus orbit of $\mathbf{X}_F(\mathbb{T})$. Its topological closure is given by*

$$V(\tau) := \overline{O(\sigma)} = \coprod_{\sigma \subseteq \tau} O(\tau).$$

The orbit closure has itself the structure of a toric variety:

$$\mathbf{X}_{\text{star}_F(\sigma)}(\mathbb{T}) = \coprod_{\tau' \in \text{star}_F(\sigma)} N'(\tau')$$

with lattice $N' = N/V_\sigma \subseteq N(\sigma)$ and fan $\text{star}_F(\sigma)$ consisting of cones $\tau' = \tau/V_\sigma$ for all cones $\tau \supseteq \sigma$ of F.

PROOF. This is the same result as in the complex case [**Ewa96**, Lemma 4.4] and could therefore be obtained via tropicalization. This is worked out in [**Pay09a**, Section 3]. □

Lemma 1.31. *Let N be a lattice and F be a fan in $N \otimes \mathbb{R}$. $\mathbf{X}_F(\mathbb{T})$ is compact if and only if $\mathbf{X}_F(\mathbb{C})$ is compact.*

PROOF. The exponential function $\exp : \mathbb{T} \to \mathbb{R}_{\geq 0}$ induces a homeomorphism between $\mathbf{X}_F(\mathbb{T})$ and $\mathbf{X}_F(\mathbb{R}_{\geq 0})$. The absolute value $|\cdot| : \mathbb{C} \to \mathbb{R}_{\geq 0}$ induces a retraction $\mathbf{X}_F(\mathbb{C}) \to \mathbf{X}_F(\mathbb{R}_{\geq 0})$. Hence $\mathbf{X}_F(\mathbb{T})$ is compact if $\mathbf{X}_F(\mathbb{C})$ is compact.
There is also a surjection $\mathbb{R}_{\geq 0} \times S^1 \to \mathbb{C}$ which induces a surjection $\mathbf{X}_F(\mathbb{R}_{\geq 0}) \times (S^1)^{\dim N} \to \mathbf{X}_F(\mathbb{C})$ (see [**Ful93**, Section 4.2]). Hence $\mathbf{X}_F(\mathbb{C})$ is compact if $\mathbf{X}_F(\mathbb{T})$ is compact. □

Corollary 1.32. *If $F \subseteq N \otimes \mathbb{R}$ is complete then $\mathbf{X}_F(\mathbb{T})$ is compact.*

PROOF. Let $\mathbf{X}_F(\mathbb{C})$ be the complex toric variety corresponding to F. It is compact in the Euclidean topology if F is complete ([**Ful93**, Prop. 2.4]). We will provide a direct proof of this result in Lemma 3.24. □

Theorem 1.33. *Let F be the normal fan of a lattice polytope P. Then there is a homeomorphism $\mu : \mathbf{X}_F(\mathbb{T}) \to P$ such that $\mu|_{O(\sigma)}$ is an analytic homeomorphism to the interior of the face of P normal to σ.*

PROOF. The topological semigroups $(\mathbb{T}, +)$ and $(\mathbb{R}_{\geq 0}, \cdot)$ are isomorphic via the map $\exp : \mathbb{T} \to \mathbb{R}_{\geq 0}$. This induces a homeomorphism between $\mathbf{X}_F(\mathbb{T})$ and $\mathbf{X}_F(\mathbb{R}_{\geq 0})$ which respects torus orbits.

If F is the normal fan of a polytope P, then $\mathbf{X}_F(\mathbb{R}_{\geq 0})$ is homeomorphic to P and respects torus orbits as stated ([**Ful93**, Prop. 4.2]). □

Remark 1.34. The relationship between the tropicalization of a projective toric variety and the moment map of that variety to a polytope is discussed in more detail in [**Pay09a**, Remark 3.3].

We will also use a representation of $\mathbf{X}_F(\mathbb{T})$ as a quotient of an open affine set by a torus action, similarly to the construction in [**Cox95**].

Definition 1.35 (Toric Variety as Global Quotient). Let F be a rational fan with lattice N and let $\rho_1, \ldots, \rho_r$ be the rays of F. Let $v_\rho \in N$ be the unique generator of $\rho \cap N$.

We define $N' := \mathbb{Z}^{F^{(1)}}$ and consider the fan $F' \subseteq N' \otimes \mathbb{R}$ defined as follows: For every cone $\sigma \in F$ there is a cone $\sigma' \in F'$ with $\sigma' := \mathrm{pos}(e_\rho : \rho \in \sigma^{(1)})$ where $(e_\rho)_{\rho \in F^{(1)}}$ is the standard basis of $\mathbb{R}^{F^{(1)}}$.

It is a subfan of the following fan E' which describes the toric variety $\mathbb{T}^{F^{(1)}}$. For any subset $S \subseteq F^{(1)}$ the set $\mathrm{pos}(e_\rho : \rho \in S)$ is a cone of E'. This means $\mathbf{X}_{F'}$ is a subvariety of the affine toric variety $\mathbf{X}_{E'} = \mathbb{T}^{F^{(1)}}$.

For each cone $\sigma \in E'$ we define the linear functions

$$x^\sigma := \sum_{\rho \in \sigma^{(1)}} x_\rho$$

and

$$x^{\hat\sigma} := \sum_{\rho \notin \sigma^{(1)}} x_\rho$$

where x_ρ is the coordinate function corresponding to the ray ρ in the vector space $\mathbb{R}^{F^{(1)}}$.

Every cone σ' of $E' \setminus F'$ corresponds to a set $Z(\sigma') := \left\{ x \in \mathbb{T}^{F^{(1)}} \mid x^{\sigma'} = -\infty \right\}$. The union $\bigcup_{\sigma' \in E' \setminus F'} Z(\sigma')$ is equal to

$$Z_F := \left\{ x \in \mathbb{T}^{F^{(1)}} \mid x^{\hat\sigma} = -\infty \, \forall \sigma \in F \right\}.$$

Let us finally consider the linear map $p_F : \mathbb{R}^{F^{(1)}} \to N$ defined via $e_\rho \mapsto v_\rho$. We define G_F as the kernel of p_F. By construction, $p_F : F' \to F$ is a surjective map of fans and hence determines a surjection $p_F : \mathbb{T}^{F^{(1)}} \setminus Z_F \to \mathbf{X}_F$.

Theorem 1.36. *Let F be a simplicial fan. $\mathbf{X}_F(\mathbb{T})$ is homeomorphic to the topological quotient $\mathbf{X}_{F'}(\mathbb{T})/G_F$ and the homeomorphism respects torus actions.*

PROOF. This is true for the complex case [**Cox95**, Theorem 2.1] and therefore also true for toric varieties over $\mathbb{C}\{\{t^\mathbb{R}\}\}$. Hence we get the result (which uses a lot of algebra) via tropicalization. □

This quotient construction allows us to equip arbitrary simplicial toric varieties with a homogeneous coordinate ring. In the complex case the description
$$\mathbf{X}_F(\mathbf{C}) = (\mathbf{C}^{F^{(1)}} \setminus Z_F)/G_F(\mathbf{C})$$
leads to an open cover with sets $U'_\sigma/G_F = \operatorname{Spec}\left(\mathbf{C}[F^{(1)}]_{x^{\hat\sigma}}\right)^{G_F}$ for $\sigma \in F$. The ring $\mathbf{C}[F^{(1)}]$ is called the homogeneous coordinate ring of $\mathbf{X}_F(\mathbf{C})$. We will now repeat this construction tropically.

Definition 1.37 (Homogeneous Coordinates). We consider the tropical polynomial ring $A_F := \mathbb{T}\left[x_\rho : \rho \in F^{(1)}\right]$. It is the semiring equivalent to the usual polynomial ring with variables x_ρ indexed by the rays of F. That means that set-theoretically elements are functions from the set of multi-indices $\mathbb{N}^{F^{(1)}}$ to the space of coefficients $\mathbb{T}$ that are almost always the neutral element of $\mathbb{T}$.

Each tropical polynomial $f = \bigoplus a_I \odot x^I$ can be evaluated at a point $p \in \mathbb{T}^{F^1}$ leading to the assignment $p \mapsto f(p) = \bigoplus a_I \odot p^I = \max(a_I + I \cdot p)$. You should think of tropical polynomials as the set of all such functions (even though the correspondence between tropical polynomials and the functions coming from tropical polynomials is not one-to-one).

We define the inclusion $i : N^\vee \to \mathbb{Z}^{F^{(1)}}, m \mapsto \sum (mv_\rho)e_\rho$ and define the class group of F

$$\operatorname{Cl}(F) := \mathbb{Z}^{F^{(1)}}/N^\vee.$$

We give A_F a $\operatorname{Cl}(F)$ grading via $\deg x_\rho := [e_\rho] \in \operatorname{Cl}(F)$.

Remark 1.38. A map $f : \operatorname{Cl}(F) \to \mathbb{R}$ is a map $\mathbb{Z}^{F^{(1)}} \to \mathbb{R}$ such that $f(N^\vee) = 0$, hence
$$\begin{aligned}\hom(\operatorname{Cl}(F), \mathbb{R}) &= \left\{x \in \mathbb{R}^{F^{(1)}} \mid \sum \langle m, v_\rho\rangle x_\rho = 0 \text{ for all } m \in N^\vee\right\} \\ &= \ker p_F \\ &= G_F.\end{aligned}$$

We obviously have a bilinear map $\operatorname{Cl}(F) \times G_F \to \mathbb{R}$ coming from the pairing
$$\operatorname{Cl}(F) \times \hom(\operatorname{Cl}(F), \mathbb{R}) \to \mathbb{R}.$$

If $Y = \mathbf{X}_F(\mathbf{C})$ is the complex toric variety defined by F, then $\operatorname{Cl}(F)$ is isomorphic to the Chow group of codimension one $A_{n-1}(Y)$.

Theorem 1.39. *Let $\mathbf{X}_F(\mathbb{T})$ be a tropical toric variety and f a homogeneous tropical polynomial from its coordinate ring A_F. Let $g \in G_F$. Then $f(g + x) = \deg(f) \cdot g + f(x)$.*

In particular: If f, h are homogeneous of the same degree, then $f(x) - h(x)$ defines a function $\mathbf{X}_F(\mathbb{T}) \setminus h^{-1}(\{-\infty\}) \to \mathbb{T}$.

PROOF. We have $f(x) = \max a_i + D_i \cdot x$ with $\deg x^{D_i} = \deg x^{D_j}$ for all $i \neq j$. Then
$$\begin{aligned} f(g+x) &= \max\{a_i + D_i \cdot (g+x)\} \\ &= \max\{a_i + D_i \cdot x + D_i \cdot g\} \\ &= \max\{a_i + D_i \cdot x + \deg(f) \cdot g\} \\ &= \max\{a_i + D_i \cdot x\} + \deg(f) \cdot g. \end{aligned}$$
□

Example 1.40. Let F be the fan of projective n-space. It has edges $-e_1, \ldots, -e_n$ and $e_0 = \sum e_i$. This means $A_F = \mathbb{T}[x_0, \ldots, x_n]$. We have
$$Z_F = \{x \in \mathbb{T}^n \mid x_i = -\infty \text{ for } i = 0, \ldots, n\} = \{-\infty, \ldots, -\infty\}$$
and $\mathrm{Cl}(F) = \mathbb{Z}$ in the exact sequence $0 \to \mathbb{Z}^n \xrightarrow{p_F} \mathbb{Z}^{n+1} \to \mathrm{Cl}(F) \to 0$.
This leads to $G_F = \mathbb{R}$ with embedding
$$\begin{aligned} G_F &= \{x \mid x_0 - x_i = 0 \text{ for } i = 1, \ldots, n\} \subseteq \mathbb{R}^{n+1} \\ &= \mathbb{R}(1, \ldots, 1). \end{aligned}$$
Therefore, we view $\mathbb{TP}^n := \mathbb{T}^n \setminus \{-\infty, \ldots, -\infty\}/\mathbb{R}(1, \ldots, 1)$ as tropical projective space.
This fan is the normal fan of the simplex $\mathrm{conv}(0, e_1, \ldots, e_n)$. Hence $\mathbb{TP}^n$ is isomorphic as cell complex to an n-simplex.

Definition 1.41. We will later need lots of projective spaces, therefore we introduce the abbreviations $\mathbb{P}(n) = \mathbb{P}^{n-1}$ and $\mathbb{P}\binom{n}{2} = \mathbb{P}^{\binom{n}{2}-1}$ for projective spaces over the tropical numbers and over algebraically closed fields.

Definition 1.42 (Linear map). Let $F \subseteq V$, $G \subseteq V'$ be two fans, $A \subseteq \mathbf{X}_F(\mathbb{T})$ and $B \subseteq \mathbf{X}_G(\mathbb{T})$ arbitrary non-empty subsets. Let $L : A \to B$ be a map. We say that L is a linear map if there are subfans F' of F and G' of G such that $A \subseteq \mathbf{X}_{F'}(\mathbb{T}) \subseteq \mathbf{X}_F(\mathbb{T})$ and $B \subseteq \mathbf{X}_{G'}(\mathbb{T}) \subseteq \mathbf{X}_G(\mathbb{T})$ and there is a map of fans $l : F' \to G'$ such that $L|_A = l|_A$.

Example 1.43. We consider the set
$$A = \{[(x, y, z)] \in \mathbb{TP}^2 \mid \max(x, y, z) \text{ is attained twice}\}.$$
Then $f : A \to \mathbb{TP}^1$, $[(x, y, z)] \mapsto [(x, y)]$ is a linear map (Figure 4 on the next page).

Definition 1.44. Let F be a complete smooth fan. To every ray $\rho \in F^{(1)}$ corresponds a subset $D_\rho := V(\rho) \subseteq \mathbf{X}_F(\mathbb{T})$ called a boundary divisor.

We understand D_ρ by looking at it in the charts U_σ:

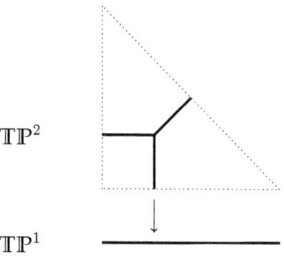

FIGURE 4. A linear map from a subset of $\mathbb{P}^2$ to $\mathbb{P}^1$.

If $\rho \in \sigma^{(1)}$ then $U_\sigma \cap D_\rho = \{x \in U_\sigma \mid x(v_\rho^\vee) = -\infty\} \cong \mathbb{T}^{n-1} \times \{-\infty\} \subseteq \mathbb{T}^n$. D_ρ is one of the n boundary faces of $U_\sigma \cong \mathbb{T}^n$.

If $\rho \notin \sigma^{(1)}$, then $U_\sigma \cap D_\rho = \emptyset$.

In homogeneous coordinates, we have $D_\rho = \{x \in \mathbf{X}_F(\mathbb{T}) \mid x_\rho = -\infty\}$.

As in the classical case, D_ρ is itself a toric variety: in the lattice $N/\operatorname{span}(v_\rho)$ the fan consists of all cones $\sigma/\operatorname{span}(v_\rho)$ for $\rho \subseteq \sigma \in F$.

If $X = \mathbf{X}_F(\mathbb{T})$ is a toric variety, then $\partial X := X \setminus V$ is the union of all boundary divisors:

$$\partial X = \bigcup_{\rho \in F^{(1)}} D_\rho.$$

CHAPTER 2

Tropical Intersection Theory

In this chapter we will introduce an intersection theory for polyhedral complexes inside tropical toric varieties that mirrors the intersection theory for algebraic cycles in complex toric varieties. We begin with a review of the tropical intersection theory established so far, then extend it to tropical toric varieties and finally relate it to the intersection theory in non-Archimedean toric varieties.

1. Tropical Polyhedral Complexes

Definition 2.1. Let σ be a polyhedral cone and τ a facet of σ, then $u_{\sigma/\tau}$ is defined to be the unique positive generator of $(N \cap \mathrm{span}(\sigma))/(N \cap \mathrm{span}(\tau)) \cong \mathbb{Z}$ (oriented such that points of $\sigma \cap N$ are positive). A vector $v \in N$ is called a primitive normal vector of σ over τ if $u_{\sigma/\tau} = v + \mathrm{span}(\tau)$.

If $P \subseteq V$ is a polyhedron and σ a facet of P then we say that $v \in N$ is a primitive normal vector of P with respect to σ if $(v, 1) \in N \times \mathbb{Z}$ is a primitive normal vector of $\mathrm{pos}(P \times \{1\})$ over $\mathrm{pos}(\sigma \times \{1\})$.

Definition 2.2. A weighted complex is a rational polyhedral complex C in a vector space V such that all inclusion-maximal polyhedra have the same dimension d and there is a weight function $w : C^{(d)} \to \mathbb{Z}$.

A weighted complex C of dimension d is balanced, if for all $\tau \in C^{(d-1)}$:

$$\sum_{\sigma > \tau} w(\sigma) u_{\sigma/\tau} = 0 \in V/\mathrm{span}(\tau).$$

Definition 2.3. Let C, D be two k-dimensional weighted polyhedral complexes in a vector space V. We call C a refinement of D if the following holds:

(1) $|C| = |D|$.
(2) Every maximal cone σ of C lies in a maximal cone τ of D and $w_C(\sigma) = w_D(\tau)$.

Two polyhedral complexes C, D are equivalent if they have a common refinement, i.e. there is a polyhedral complex E such that E is a refinement of both C and D.

Lemma 2.4. [AR09, Constr. 2.13] *Let C, D be weighted polyhedral complexes of the same dimension. Then there are refinements C' of C and D' of D such that $C' \cup D'$ is a pure-dimensional polyhedral complex. We turn it into a weighted complex by setting $w_{C+D}(\sigma) = w_C(\sigma) + w_D(\sigma)$.*

Lemma 2.5. [AR09, Lemma 2.11] *Let C be a weighted polyhedral complex and D a refinement of C. Then C is balanced if and only if D is balanced.*

Theorem 2.6. [AR09, Lemma 2.14] *The classes under refinement of k-dimensional balanced polyhedral complexes in T form an Abelian group $Z_k(T)$.*

Definition 2.7 (Tropical Polyhedral Complex). A tropical polyhedral complex is the class under refinements of a balanced polyhedral complex.

Remark 2.8. Usually we do not care about the polyhedral structure of our complexes, that is why we consider different complexes as equal if they have a common refinement. One example of this are tropicalizations of algebraic varieties. These always have a structure of a balanced polyhedral complex ([BJS+07, Theorem 2.9]), but there is no unique or canonical structure (see Example 2.9).

The polyhedral structure is important if we consider fans of toric varieties. For example all complete fans of the same dimension are equal up to common refinements.

Example 2.9. Let $V = \mathbb{R}^4$ and $L_1 = \mathbb{R}^2 \times \{0\}$, $L_2 = \{0\} \times \mathbb{R}^2$.

We consider L_1 and L_2 as weighted polyhedral complexes with weight one. The sum $L_1 + L_2$ does not come with a canonical polyhedral structure on $L_1 \cup L_2$. The origin lies in the intersection $L_1 \cap L_2$, but it is not a face of either. We need to choose a complete fan in L_1 and another complete fan in L_2 to make $L_1 \cup L_2$ into a polyhedral complex.

Definition 2.10. Let $[F]$ be a class of fans under refinements. $[F]$ is called hübsch if there is a (necessarily unique) fan G such that $[F]$ is the class of refinements of G. In other words: the set $|F|$ has a unique coarsest fan structure.

Remark 2.11. We have already seen an example of a fan that is not hübsch, Example 2.9. Our main examples of hübsch fans will be Bergman fans and the space of trees $M_{0,n}$. Hübsch fans are important for tropical compactifications [LQ09].

2. Stable Intersections of Tropical Complexes

Definition 2.12 (Tropical Rational Function). Let N be a lattice and T the vector space $N \otimes \mathbb{R}$. A tropical rational function on T is a continuous piecewise affine linear function $r : T \to \mathbb{R}$ satisfying the following conditions

(1) There is a finite cover $T = \bigcup P_i$ of T with polyhedra with rational slopes such that r is affine linear on each P_i.
(2) Let P be any polyhedron such that r is affine linear on P. Then there is a (necessarily unique) vector $r_P \in (N \cap \text{span}(P))^\vee$ and a number $r_{P,0}$ such that $r(x) = r_P \cdot x + r_{P,0}$ for all x in P.

The set of all tropical rational functions on T is an Abelian group, it contains the subgroup $N^\vee$ of all tropical rational functions that are linear everywhere and the subgroup $\mathbb{R}$ of constant functions $T \to \mathbb{R}$.

Two tropical rational functions are considered equivalent if their difference lies in $\mathbb{R}$. Most of the time we will only consider tropical rational functions up to this equivalence and therefore set
$$\text{Rat}(T) := \{\text{tropical rational functions on } T\}/\mathbb{R}.$$
The linear functions from $N^\vee$ form a subgroup of $\text{Rat}(X)$ since $N^\vee/\mathbb{R} \cong N^\vee$.

Remark 2.13. The difference of two tropical polynomials is such a piecewise affine linear function.

Definition 2.14 (Tropical Intersection Product). Let $[r]$ be an element of $\text{Rat}(T)$ and C be a k-dimensional tropical polyhedral complex. We define a $(k-1)$-dimensional weighted complex $[r] \cdot C$ as follows:

Choose a refinement of C such that r is affine linear on every cell of C. Maximal cells of $[r] \cdot C$ are codimension one cells of C, for such a cell Q, the weight of Q in $[r] \cdot C$ is given by the formula
$$\text{ord}_Q(r) := \sum_{P > Q} w(P) r_P(v_{P/Q}) - r_Q\left(\sum w(P) v_{P/Q}\right)$$
for any choice of representative $[r]$ and primitive normal vectors $v_{P/Q}$ representing $u_{P/Q}$.

Theorem 2.15. *Let r, r' be a tropical rational functions and C be a tropical polyhedral complex.*

(1) $[r] \cdot C$ *is a well-defined tropical polyhedral complex.*
(2) $([r] + [r']) \cdot C = [r] \cdot C + [r'] \cdot C$ *and* $[r] \cdot (C + C') = [r] \cdot C + [r] \cdot C'$.
(3) $[r] \cdot [r'] \cdot C = [r'] \cdot [r] \cdot C$.

PROOF. [AR09, Prop. 3.7, Prop. 6.7]. □

Definition 2.16. Let X be a polyhedral complex in $T = N \otimes \mathbb{R}$ and X' a polyhedral complex in $T' = N' \otimes \mathbb{R}$. A morphism $X \to Y$ is a linear map $T \to T'$ that maps $|X|$ to $|Y|$ and N to N'.

Definition 2.17 (Push-Forward). Let X be a polyhedral complex in $T = N \otimes \mathbb{R}$ and X' a polyhedral complex in $T' = N' \otimes \mathbb{R}$. Let $f : X \to X'$ be a morphism. Then the push-forward

f_*X is the weighted polyhedral complex with cells

$$\{f(P) \mid P \in X \text{ contained in a maximal cone on which } f \text{ is injective}\}$$

and weights

$$w(Q) := \sum_{f(P)=Q} w(P) \left[N'_Q : f(N_P)\right]$$

Definition 2.18 (Pull-Back). Let $r : T \to \mathbb{R}$ be a tropical rational function and $f : S \to T$ a morphism. The pull-back $f^*r : S \to \mathbb{R}$ is defined as the piecewise affine linear function $x \mapsto r(f(x))$.

Theorem 2.19 (Projection Formula). *Let $f : C \to D$ be a morphism, E a tropical polyhedral complex on C and r a rational function on D. Then*

$$[r] \cdot (f_*E) = f_*([f^*r] \cdot E)$$

PROOF. [**AR09**, Prop. 4.8, Prop. 7.7] □

Definition 2.20 (Stable Intersection). Let C and D be balanced polyhedral complexes of codimensions p and q in the vector space T. we define a balanced polyhedral complex $C \cdot D$ of codimension $p + q$ via $C \cdot D = \mathrm{pr}_*([\Delta] \cdot C \times D)$ where $[\Delta]$ is a product of tropical rational functions describing the diagonal in $T \times T$ and $\mathrm{pr} : T \times T \to T$ is the projection onto the first factor.

Remark 2.21. The name stable intersection was originally used in [**RGST05**] for intersections of generic tropical curves in the plane. A more elaborate theory was suggested in [**Mik06b**] and developed in [**AR09, AR08**].

A drawback of the theory is that the intersection of tropical polyhedral complexes of dimensions k and h in an ambient vector space T of dimension n will always be of dimension $n - k - h$, even if there is a subvariety E of T containing both cycles (e.g. one cannot intersect two curves inside a tropical hypersurface of $\mathbb{R}^3$).

One approach to define an intersection product in the ambient variety E is to express the diagonal in $E \times E$ as a product of tropical rational functions. This is an area of active research (see for example [**FR10**]).

Theorem 2.22. *Let C, D and E be tropical polyhedral complexes in T. Let $[r] : T \to \mathbb{R}$ be a tropical rational function. Then*

(1) $T \cdot C = C$.
(2) $C \cdot D = D \cdot C$.
(3) $C \cdot (D + E) = C \cdot D + C \cdot E$ *if D and E are of the same dimension.*
(4) $([r] \cdot C) \cdot D = [r] \cdot (C \cdot D)$.
(5) $(C \cdot D) \cdot E = C \cdot (D \cdot E)$.

PROOF. [AR09, Cor. 9.5, Lemma 9.7, Theorem 9.10] □

Remark 2.23. We have $|C \cdot D| \subseteq |C| \cap |D|$. Under suitable conditions (transversal intersection) a maximal cell R of $C \cdot D$ is the intersection $R = P \cap Q$ of maximal cells P of C and Q of D with the weight
$$w(R) = w(P) \cdot w(Q) \cdot [N : N \cap \operatorname{span} P + N \cap \operatorname{span} Q]$$
[Rau09, Cor. 1.5.16].

A remarkable feature of this intersection product is that we can always perform the intersection and never have to pass to classes modulo rational equivalence unlike the classical intersection theory. The reason for this turns out to be that tropical fans actually represent classes of complex cycles modulo rational equivalence.

Definition 2.24. Let N be a lattice and F a complete unimodular fan in $T = N \otimes \mathbb{R}$. A balanced polyhedral fan with support in $F^{(k)}$ is called a Minkowski weight of dimension k.

Minkowski weights of dimension k form a group $\mathrm{MW}_k(F)$ and all Minkowski weights form a graded ring $\mathrm{MW}_*(F)$ with multiplication the stable intersection of fans in $\mathbb{R}^n$.

Remark 2.25. The ring of Minkowski weights was introduced in [**FS97**], where it was given an explicit ring structure. It was shown in [**Kat09a**] and [**Rau09**] that this is actually the multiplication of tropical intersection theory as defined in [**AR09**, Def 9.3] (Definition 2.20 in this work).

Remark 2.26. The group of Minkowski weights $\mathrm{MW}_{n-k}(F)$ is naturally dual to the Chow group $A_k(\mathbf{X}_F(\mathbb{C}))$. Let $[a]$ be an element of $A_k(\mathbf{X}_F(\mathbb{C}))$, it is represented by a sum $a = \sum a_\sigma V(\sigma)$ of k-dimensional orbit closures (hence all cones σ are of codimension k). Let m be a Minkowski weight, it consists of a number $m(\sigma)$ for every $n - k$-dimensional cone σ of F. The pairing $\mathrm{MW}_{n-k}(F) \times A_k(\mathbf{X}_F(\mathbb{C})) \to \mathbb{Z}$ is simply given by $m \cdot [a] = \sum a_\sigma m(\sigma)$. The balancing condition on m guarantees that is independent of the chosen representative for the class $[a]$.

Let $[b]$ be an element from $A_{n-k}(\mathbf{X}_F(\mathbb{C}))$ with $b = \sum b_\tau V(\tau)$. Via a suitable choice of representatives of $[a]$ and $[b]$ we can achieve that for every cone σ with $a_\sigma \neq 0$ and every cone τ with $b_\tau \neq 0$ the intersection $V(\tau) \cap V(\sigma)$ is either a point represented by $V(\operatorname{pos}(\tau \cup \sigma))$ or empty. All these points are equivalent and this defines a map $A_{n-k}(\mathbf{X}_F(\mathbb{C})) \to \hom(A_k(\mathbf{X}_F(\mathbb{C})), \mathbb{Z})$ with the pairing
$$[b] \cdot [a] = \sum_{\operatorname{pos}(\sigma \cup \tau) \in F^{(n)}} a_\sigma b_\tau.$$
It turns out that this map is an isomorphism.

Theorem 2.27. *Let F be a complete unimodular fan. Then the group of Minkowski weights $\mathrm{MW}_k(F)$ is isomorphic to the Chow group $A_k(\mathbf{X}_F(\mathbb{C}))$.*

PROOF. Let n be the dimension of $\mathbf{X}_F(\mathbb{C})$. Minkowski weights of dimension k are canonically isomorphic to the dual of Chow groups $\hom(A_{n-k}(\mathbf{X}_F(\mathbb{C})), \mathbb{Z})$ [FS97, Prop. 1.4]. By Poincaré-duality, $\hom(A_{n-k}(\mathbf{X}_F(\mathbb{C})), \mathbb{Z})$ is isomorphic to $A_k(\mathbf{X}_F(\mathbb{C}))$. Note that [FS97] indexes Chow groups by codimension whereas we index them by dimension. □

3. Intersection Theory on Tropical Toric Varieties

Definition 2.28 (Tropical Cycle). Let $X = \mathbf{X}_F(\mathbb{T})$ be a tropical toric variety and $X = \coprod O(\sigma)$ be its decomposition into torus orbits. A k-cycle on X is a collection $C = (C_\sigma)_{\sigma \in F}$ of tropical polyhedral complexes of dimension k in each $O(\sigma)$ with $\dim O(\sigma) \geq k$.

The group of all k-cycles on X is denoted $Z_k(X)$.

Definition 2.29. Let N be a lattice and F a fan in $T = N \otimes \mathbb{R}$.

- Let σ be a cone of F. The lattice $N(\sigma)$ is defined as $N/\text{span}(\sigma)$. It is the lattice of the toric subvariety $V(\sigma)$ of $\mathbf{X}_F$.
- Let P be a polyhedron in T. We define a lattice N_P as $N \cap \text{span}(P)$. It is the lattice generated by the lattice points of P.

Definition 2.30 (Tropical Rational Function). A rational function on a tropical toric variety is a tropical rational function on the torus of that variety.

Definition 2.31. Let $r : T \to \mathbb{R}$ be a tropical rational function on $X = \mathbf{X}_F(\mathbb{T})$ and $\rho \in F^{(1)}$ be a ray. Let $P \subseteq T$ be a polyhedron containing ρ in its recession cone such that r is affine linear on P. The multiplicity or order of vanishing of r along ρ is defined as $\text{ord}_\rho(r) = r_P(-v_\rho)$.

Lemma 2.32. *For each ray ρ the map $\text{ord}_\rho : \text{Rat}(X) \to \mathbb{Z}$ is a well-defined group homomorphism.*

PROOF. We first show that it is well-defined. Let $P \subseteq T$ be a polyhedron containing ρ in its recession cone such that r is affine linear on P. Let $P' \subseteq T$ be another polyhedron containing ρ in its recession cone such that r is affine linear on P'.

Let us assume that P and P' are adjacent. Let F be the intersection $P \cap P'$. Since both P and P' contain the ray ρ in their recession cone, this is also true for F. We then see $r_P v + r_{P,0} = r_F v + r_{F,0} = r_{P'} v + r_{P',0}$ for all $v \in F$. That means $r_P - r'_P \in F^\perp$, in particular $r_P(-v_\rho) = r_F(-v_\rho) = r'_P(-v_\rho)$.

If P and P' are not adjacent, we can find a sequence $P = P_0, P_1, \ldots, P_k = P'$ of adjacent polytopes that all contain the ray ρ in their recession cones.

The map ord_ρ is by definition linear and zero on constant functions. □

Definition 2.33. Let r be a tropical rational function and $O(\sigma)$ a torus orbit. We say r restricts to $O(\sigma)$ if the assignment

$$z \mapsto r^\sigma(z) = \lim_{\substack{x \in T \\ x \to z}} r(x)$$

defines a tropical rational function $O(\sigma) \to \mathbb{R}$.

Remark 2.34. This is the case if and only if $\mathrm{ord}_\rho(r) = 0$ for all $\rho \in \sigma$.

Definition 2.35 (Tropical Cartier Divisor). Let $X = \mathbf{X}_F$ be an n-dimensional tropical toric variety with torus T and C a k-cycle on X. A Cartier divisor on C is a finite family $\varphi = (U_\alpha, r_\alpha)$ of pairs of open subsets U_α of $\overline{|C|}$ and tropical rational functions r_α on X satisfying the following conditions:

- The union of all U_α covers $\overline{|C|}$.
- For every component C_σ of C in $O(\sigma)$ and every chart U_α such that $U_\alpha \cap C_\sigma \neq \emptyset$ the function r_α must restrict to $O(\sigma)$.
- For every component C_σ of C in $O(\sigma)$ and all charts U_α, U_β such that $U_\alpha \cap U_\beta \cap |C_\sigma| \neq \emptyset$ there is an affine linear tropical rational function d such that $r_\alpha^\sigma(x) - r_\beta^\sigma(x) = d(x)$ for all $x \in U_\alpha \cap U_\beta \cap |C_\sigma|$ and d extends to a continuous function $d : U_\alpha \cap U_\beta \to \mathbb{R}$.

Two Cartier divisors $\varphi = (U_\alpha, r_\alpha)$, $\psi = (W_\beta, s_\beta)$ are considered equal if

- For every component C_σ of C in $O(\sigma)$ and all charts U_α, V_β such that $U_\alpha \cap V_\beta \cap |C_\sigma| \neq \emptyset$ there is an affine linear tropical rational function d such that $r_\alpha^\sigma(x) - s_\beta^\sigma(x) = d(x)$ for all $x \in U_\alpha \cap V_\beta \cap |C_\sigma|$ and d extends to a continuous function $d : U_\alpha \cap V_\beta \to \mathbb{R}$

Cartier divisors form an Abelian group $\mathrm{Cart}(X)$ under chart-wise addition of tropical rational functions. Tropical rational functions are included as the subgroup of Cartier divisors that have the same function in every chart.

Remark 2.36. If $C = \mathbf{X}_F(\mathbb{T})$ then these conditions simplify to:

- The union of all U_α covers $\mathbf{X}_F(\mathbb{T})$.
- For all charts U_α, U_β such that $U_\alpha \cap U_\beta \neq \emptyset$ there is an affine linear tropical rational function d such that $r_\alpha(x) - r_\beta(x) = d(x)$ for all $x \in U_\alpha \cap U_\beta \cap T$ and d extends to a continuous function $d : U_\alpha \cap U_\beta \to \mathbb{R}$.

Example 2.37. The easiest way to construct Cartier divisors that are not rational functions is via homogeneous tropical polynomials. Let F be a complete unimodular fan and f a homogeneous polynomial from A_F. On every maximal chart $U_\sigma \cong \mathbb{T}^n$ there is a tropical polynomial f_σ, the dehomogenization of f, obtained by substituting 0 into all variables x_ρ such that the ray ρ is not contained in σ. The collection (U_σ, f_σ) then constitutes a Cartier divisor.

Definition 2.38. Let $\varphi = (U_\alpha, r_\alpha)$ be a Cartier divisor on $X = \mathbf{X}_F(\mathbb{T})$ and $\rho \in F^{(1)}$ be a ray. We choose a chart β containing points of $O(\rho)$ and define $\mathrm{ord}_\rho(\varphi) := \mathrm{ord}_\rho(r_\beta)$ as the multiplicity of φ along ρ.

Lemma 2.39. *This multiplicity is well-defined.*

PROOF. Let γ be a another chart that contains points of $O(\rho)$. Assume for now that there is a point z contained in $U_\beta \cap U_\gamma \cap O(\rho)$. That means $(r_\beta - r_\gamma) = mx + k$ for some $m \in N^\vee$ and $k \in \mathbb{R}$. Furthermore, the limit $\lim_{x \to z} mx + k$ exists, which is only possible if $mv_\rho = 0$. Hence $\mathrm{ord}_\rho(r_\gamma - r_\beta) = 0$.

We know that $V(\rho)$ gets covered by finitely many open U_α, which means we can find a chain of charts $U_\beta = U_{\alpha_0}, \ldots, U_{\alpha_h} = U_\gamma$ such that $U_{\alpha_i} \cap U_{\alpha_{i+1}} \cap O(\rho) \neq \emptyset$. □

Lemma 2.40. *Let φ be a Cartier divisor on X. Then there is a tropical rational function s and representative $\varphi = (U_\alpha, r_\alpha)$ with $r_\alpha - s \in N^\vee + \mathbb{R}$ for all α.*

PROOF. Choose a simply connected chart U_{α_0} containing the point $0 \in T$. Set $s_{\alpha_0} = r_{\alpha_0}$. We can (after a suitable enlargement of the atlas) now cover X with charts $U_{\alpha_0}, \cdots, U_{\alpha_h}$ such that for every i the set $\bigcup_{j=1}^i U_{\alpha_j} \cap U_{\alpha_{i+1}}$ is connected. We know that there is an $m \in N^\vee$ such that the map $x \mapsto r_{\alpha_i}(x) - r_{\alpha_{i+1}} + mx$ is locally constant and therefore constant. By induction there is also an $m' \in N^\vee$ and $\lambda \in \mathbb{R}$ such that $x \mapsto r_{\alpha_{i+1}}(x) - s_{\alpha_o} + m'x + \lambda$ is locally zero and therefore zero.

Hence the map $s : T \to \mathbb{R}$, $x \mapsto s_{\alpha_i}(x)$ for any U_{α_i} containing x is a well-defined continuous piecewise affine linear function satisfying the claim. □

Lemma 2.41. *A Cartier divisor φ on an n-dimensional complete smooth tropical toric variety $\mathbf{X}_F(\mathbb{T})$ can be represented as $\varphi = (U_\sigma, r_\sigma)_{\sigma \in F^{(n)}}$.*

PROOF. After applying Lemma 2.40 we can assume that every chart contains T. Each orbit $O(\sigma)$ consists of just one point. Each chart U_α containing the point $O(\sigma)$ must meet all orbits $O(\rho)$ for every ray ρ contained in σ. We set $r_\sigma := r_\alpha$ and see that $(U_\alpha, r_\alpha) = (U_\sigma, r_\sigma)$. □

Lemma 2.42. *A Cartier divisor on X is uniquely characterized by an element $[s]$ in $\mathrm{Rat}(X)/N^\vee$ and a collection $(a_\rho)_{\rho \in F^{(1)}}$ of integers.*

PROOF. We assume that the Cartier divisor is given as $\varphi = (U_\sigma, s_\sigma)$. If we start with a Cartier divisor $\varphi = (U_\sigma, r_\sigma)$, and choose $a_\rho = \mathrm{ord}_\varphi(\rho)$ and $s = r_\sigma$ for an arbitrary maximal cone σ. We will now show how to construct a Cartier divisor ψ from such data $((a_\rho)_{\rho \in F^{(1)}}, s)$. Let σ be a maximal cone of F. Let s be any representative of $[s]$. We define a new representative s_σ such that $\mathrm{ord}_{s_\sigma}(\rho) = a_\rho$ for all rays ρ contained in σ (this can be done since σ is unimodular, as in [**Ful93**], section 3.4]). The collection (s_σ) forms a Cartier divisor ψ. We have $\psi = \phi$ since $s_\sigma - r_\sigma$ is linear and $\mathrm{ord}_\rho(s_\sigma) = \mathrm{ord}_\rho(r_\sigma)$. □

When we want to use this representation, we write a Cartier divisor as

$$\varphi = \sum a_\rho D_\rho + [s]$$

with $a_\rho = \operatorname{ord}_\rho(\varphi)$.

We will describe how to form an intersection product of tropical Cartier divisors and cycles in a tropical toric variety. In addition to using the (usual) faces of a polyhedron we will also be using infinite faces – the intersection of torus orbits with the closure of the polyhedron in the toric variety. This will allow us later on to use a significantly broader definition of rational equivalence than [**AR08**].

Lemma 2.43. *Let P be a polyhedron in T and $\overline{P}$ the closure of P in X. Then $\overline{P} \cap O(\sigma)$ is non-empty if and only if the $\sigma \cap \operatorname{rec} P \neq \emptyset$. In that case $\overline{P} \cap O(\sigma) = P/\operatorname{span}(\sigma)$ is again a polyhedron.*

PROOF. Let v be a vector in $\sigma \cap \operatorname{rec} P$ and p a point in P. Then $x_n = p + nv$ lies in P and $x = \lim_{n \to \infty} x_n$ lies in $O(\sigma)$ hence $x \in \overline{P} \cap O(\sigma)$.

In fact, we have $\overline{P} \cap O(\sigma) = \{\lim_{n \to \infty} p + nv \mid p \in P, v \in \operatorname{rec} P \cap O(\sigma)\}$. If we look at the orbit decomposition $X = \coprod O(\tau) = \coprod T/\operatorname{span}(\tau)$ then the point $x = \lim p + nv \in X$ corresponds to the point $p + \operatorname{span}(\sigma)$ in $O(\sigma)$. Hence $\overline{P} \cap O(\sigma) = P/\operatorname{span}(\sigma)$. □

Definition 2.44. Let X be a smooth complete tropical toric variety with fan F and torus T. Let C be a $(k+1)$-dimensional balanced polyhedral complex in T. Let σ be any positive dimensional cone of F.

We define a k-dimensional weighted polyhedral complex $O(\sigma) \cdot C$ in $O(\sigma)$ as follows: For every maximal cell P of C such that $P' = \overline{P} \cap O(\sigma) \neq \emptyset$ and $1 + \dim(P') = \dim P$ the polyhedron P' is a maximal cell of $O(\sigma) \cdot C$ with the weight $w(P') = [N(\sigma)_{P'} : N_P(\sigma)] \cdot w(P)$.

Lemma 2.45. *$O(\sigma) \cdot C$ is a balanced polyhedral complex.*

PROOF. Let $Q/\operatorname{span}(\sigma)$ be a codimension one cell of $O(\sigma) \cdot C$. We write P' and Q' for quotients $P/\operatorname{span}(\sigma)$ and $Q/\operatorname{span}(\sigma)$ of polyhedra of C. We need to show $\sum_{P'>Q'} w(P')u_{P'/Q'} = 0$.

We know $\sum_{P>Q} w(P) u_{P/Q} = 0$. Furthermore, since Q contains σ in its recession cone, we know that every P with $P > Q$ also contains σ in its recession cone. Hence both sums iterate over the same index set.

Let $(v_{P/Q})_P$ be a system of primitive normal vectors for the maximal cones P surrounding Q. Let $v_1, \ldots, v_r$ be a lattice basis of N_Q. Then $v_1, \ldots, v_r, v_{P/Q}$ is a lattice basis of N_P. Let $v'_1, \ldots, v'_s$ be a lattice basis of $N(\sigma)_{Q'}$. Then $v'_1, \ldots, v'_s, v_{P'/Q'}$ is a lattice basis of $N(\sigma)_{P'}$. We therefore find that the class $u_{P/Q} + \operatorname{span} \sigma = v_{P/Q} + N_Q + \operatorname{span} \sigma$ is a multiple of $u_{P'/Q'} = v_{P'/Q'} + N(\sigma)_{Q'}$.

This factor can be expressed as $[N(\sigma)_{P'} : N_P(\sigma)]/[N(\sigma)_{Q'} : N_Q(\sigma)]$, hence we have the formula $[N(\sigma)_{P'} : N_P(\sigma)] u_{P'/Q'} = [N(\sigma)_{Q'} : N_Q(\sigma)] u_{P/Q} + \operatorname{span} \sigma$.

The balancing condition around Q' then amounts to

$$\sum_{P'>Q'} w(P') u_{P'/Q'}$$
$$= \sum_{P>Q} [N(\sigma)_{P'} : N_P(\sigma)] w(P) u_{P'/Q'}$$
$$= \sum_{P>Q} w(P) [N(\sigma)_{Q'} : N_Q(\sigma)] u_{P/Q} + \operatorname{span}(\sigma)$$
$$= [N(\sigma)_{Q'} : N_Q(\sigma)] \sum_{P>Q} w(P) u_{P/Q} + \operatorname{span}(\sigma)$$
$$= 0.$$

$\square$

Lemma 2.46. *Let P be a rational polyhedron such that $P' = \overline{P} \cap O(\sigma)$ is non-empty. Assume $1 + \dim P' = \dim P$.*

(1) *There exists a unique primitve lattice vector $v_{P/P'} \in N \cap \operatorname{rec} P \cap \sigma$.*
(2) *If $Q < P$ and $\overline{Q} \cap O(\sigma) =: Q' < P'$ then $v_{P/P'} = v_{Q/Q'}$.*

PROOF.

(1) We know that such a v exists, since $\operatorname{rec} P \cap \sigma \neq \emptyset$. Assume we have two different primitive lattice vectors v, w in $\operatorname{rec} P \cap \sigma$. Then $\dim \operatorname{pos}(v, w) = 2$ and $\operatorname{pos}(v, w) \subseteq \operatorname{rec} P, \sigma$. That means $\dim P + \operatorname{span}(\sigma) \leq \dim P - 2$. Hence $v = w$.
(2) We have $v_{Q/Q'} \in N \cap \operatorname{rec} Q \cap \sigma \subseteq N \cap \operatorname{rec} P \cap \sigma$.

$\square$

Definition 2.47 (Intersection Product). Let φ be a Cartier divisor on a $(k+1)$-cycle C. We define a k-cycle as follows:

We choose primitive normal vectors $v_{P/Q}$ and rational functions φ_P in open charts containing P. For each orbit $O(\sigma)$ with $C_\sigma \neq 0$ we get a component $E_{\sigma,\sigma}$ in $O(\sigma)$ whose cells are the codimension one cells of C_σ with weight

$$w(Q) = \sum_{P>Q} w(P) \varphi_P(v_{P/Q}) - \varphi_Q(\sum w(P) v_{P/Q}).$$

For each orbit $O(\tau)$ with $\sigma \subsetneq \tau$ we get a component $E_{\sigma,\tau}$ in $O(\tau)$ whose cells are the codimension one infinite cells of C_σ with weight

$$w(P') = w(P) [N_P(\tau) : N(\tau)_{P'}] \varphi_{P'}(v_{P/P'}).$$

The intersection product $\varphi \cdot C$ is then defined as $\varphi \cdot C = \sum_{\sigma \subseteq \tau} E_{\sigma,\tau}$.

Theorem 2.48. *Let φ be a Cartier divisor on a $(k+1)$-cycle C.*

(1) $\varphi \cdot C$ *is a well-defined cycle.*
(2) $\varphi \cdot (C + D) = \varphi \cdot C + \varphi \cdot D$ *and* $(\varphi + \psi) \cdot C = \varphi \cdot C + \psi \cdot C$.

PROOF.

(1) All components of the form $E_{\sigma,\sigma}$ produce an intersection product as in [AR09, Construction 6.4]. It is shown there that this definition is well-defined and produces a balanced polyhedral complex.

Assume we an an orbit $O(\tau)$ containing an infinite facet P' of a maximal cell P in an orbit $O(\sigma)$. Let U_α, U_β be two open sets containing P'. Then the difference $\varphi_\alpha - \varphi_\beta$ is constant along $v_{P/P'}$.

For the balancing condition with respect to $E_{\sigma,\tau}$, we pick a face Q' of codimension one and then see that $v_{P/P'} = v_{Q/Q'}$ for all $P' > Q'$. Furthermore if we pick an open set U_α containing Q' then this set must also meet all P' and all P as well as Q. Hence $\varphi_{P'}(v_{P/P'}) = \varphi_{Q'}(v_{Q/Q'})$ for all $P > Q$, the balancing condition of $E_{\sigma,\tau}$ then follows from Lemma 2.45.

(2) This follows from the definition.

$\square$

Example 2.49. Let us take $X = \mathbb{TP}^2$ as ambient toric variety. We fix a chart $\mathbb{T}^2$ and identify the torus T of X with $\mathbb{R}^2 \subseteq \mathbb{T}^2$.

Let us consider the rational function $s : \mathbb{R}^2 \to \mathbb{R}$, $(x, y) \mapsto \max(2x - 1, 2y - 1, x + y + 1, x, y, 0)$. When we treat T as a balanced polyhedral complex of weight one we can form the intersection product $s \cdot T$. This is a one-dimensional balanced polyhedral complex in T, which is supported on the locus of non-linearity of s.

Let us compute the weight on the polyhedral cell $Q = \{2y - 1 = x + y + 1 \geq 0, x, y, 2x - 1\}$ It is adjacent to the polyhedral cells $P_1 = \{2y - 1 \geq x + y + 1, 0, x, y, 2x - 1\}$ and $P_2 = \{x + y + 1 \geq 2y - 1, 0, x, y, 2x - 1\}$. which are both of weight one. Let us choose $v_{P_1/Q} = (1, 0)$ and $v_{P_2/Q} = (0, 1)$ as primitive normal vectors. They satisfy $(1, 1) = v_{P_1/Q} + v_{P_2/Q} \in \mathbb{Z}^2 \cap \operatorname{span} Q = \mathbb{Z}(1, 1)$. We now have

$$\begin{aligned} w(Q) &= s_{P_1}(v_{P_1/Q}) + s_{P_2}(v_{P_2/Q}) - s_Q(v_{P_1/Q} + v_{P_2/Q}) \\ &= (2 + 1) - 2 \\ &= 1. \end{aligned}$$

By symmetry, this means that the weight on $Q_2 = \{2x - 1 = x + y + 1 \geq 0, x, y, 2y - 1\}$ is also equal to one. The polyhedron $Q_3 = \{y = x + y + 1 \geq 0, x, 2x - 1, 2y - 1\}$ is adjacent to

$P_3 = \{y \geq x + y + 1, 0, x, 2x - 1, 2y - 1\}$ and $P_4 = \{x + y + 1 \geq 0, x, y, 2x - 1, 2y - 1\}$. We choose $v_{P_3/Q_3} = (-1, 0)$ and $v_{P_4/Q_3} = (1, 0)$. We can then compute

$$\begin{aligned} w(Q_3) &= s_{P_3}(v_{P_3/Q_3}) + s_{P_4}(v_{P_4/Q_3}) - s_{Q_3}(v_{P_3/Q_3} + v_{P_4/Q_3}) \\ &= (0 + 1) - 0 \\ &= 1. \end{aligned}$$

By similar computations (more examples can be found in [**AR09**]), all weights are equal to one.

If we considered the intersection product $s \cdot \mathbb{TP}^2$, we would get additional weights on the boundary divisors of $\mathbb{TP}^2$. We want to form a Cartier divisor φ on $\mathbb{TP}^2$ such that $\varphi \cdot \mathbb{TP}^2 = s \cdot T$, i.e. $\varphi = 0 \cdot D_0 + 0 \cdot D_1 + 0 \cdot D_2 + [s]$. We can do this by the standard cover U_0, U_1, U_2 on $\mathbb{TP}^2$ with $\varphi_0 = s$, $\varphi_1 = s - 2x$ and $\varphi_2 = s - 2y$. This equals the dehomogenization of the tropical polynomial $2x - 1 \oplus 2y - 1 \oplus 2z \oplus x + y + 1 \oplus x + z \oplus y + z$ on the respective charts.

Let us now consider the rational function $r : \mathbb{R}^2 \to \mathbb{R}$, $(x, y) \mapsto \max(0, -x, -y)$. If we form the intersection product $r \cdot (s \cdot T)$ in T then we get three possible intersection points (points of $s \cdot \mathbb{R}^2$ where r is not linear). The point $p_1 = (0, 2)$ is adjacent to $Q_a = \text{conv}\{(-1, 1), (0, 2)\}$ and $Q_b = (0, 2) + \mathbb{R}_{\geq 0}(1, 1)$ (they form a subdivision of Q_1). Our primitive normal vectors are $v_{Q_a/p_1} = (-1, -1)$ and $v_{Q_b/p_1} = (1, 1)$. thus we get

$$\begin{aligned} w(p_1) &= r_{Q_a}(v_{Q_a/p_1}) + r_{Q_b}(v_{Q_b/p_1}) - r_{p_1}(v_{Q_a/p_1} + v_{Q_b/p_1}) \\ &= (1 + 0) - 0 \\ &= 1. \end{aligned}$$

By symmetry we see that the weight of $p_2 = (2, 0)$ must also be one.

Hence we have three intersection points with a combined multiplicity of four (see Figure 5 on the facing page).

However, as r is a rational function, we expect the intersection product $r \cdot (\varphi \cdot \mathbb{TP}^2)$ in $\mathbb{TP}^2$ to have a combined multiplicity of zero. The curve $\varphi \cdot \mathbb{TP}^2$ has six boundary points in $\mathbb{TP}^2 \setminus \mathbb{R}^2$. They are $q_1 = [(0, -\infty, 0)]$, $q_2 = [(0, -\infty, 1)]$ in D_1, $q_3 = [(0, 0, -\infty)]$, $q_4 = [(0, 1, -\infty)]$ in D_2 and $q_5 = [(-\infty, -1, 1)]$, $q_6 = [(\infty, 1, -1)]$ in D_0.

The point q_1 is an infinite face of $Q_1 = (-1, 0) + \mathbb{R}_{\geq 0}(-1, 0)$. Hence v_{Q_1/q_1} must be equal to $(-1, 0)$ as the recession fan of Q_1 is one-dimensional. We also see that all involved lattice indices are one as the corresponding lattices are zero-dimensional. Hence

$$\begin{aligned} w(q_1) &= w(Q_1) \left[(\mathbb{Z}^2/(-1,0)\mathbb{Z}) \cap \{0\} : (\mathbb{Z}^2 \cap (-1,0)\mathbb{Z}) / (-1,0)\mathbb{Z} \right] r_{Q_1}(v_{Q_1/q_1}) \\ &= 1 \cdot 1 \cdot (-1) \\ &= -1 \end{aligned}$$

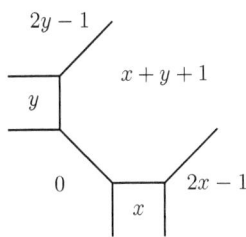
(a) The intersection $s \cdot \mathbb{R}^2$. All weights are equal to one.

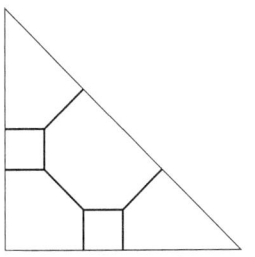
(b) The intersection $\varphi \cdot \mathbb{TP}^2$. All weights are equal to one.

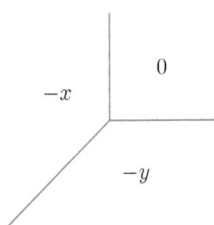
(c) The intersection $r \cdot \mathbb{R}^2$. All weights are equal to one.

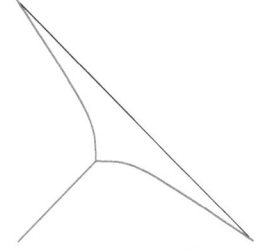
(d) The intersection $r \cdot \mathbb{TP}^2$. All weights in the interior are one, weights of the indicated boundary divisors are negative one.

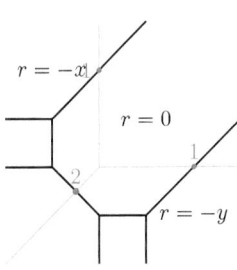
(e) The intersection $r \cdot (s \cdot \mathbb{R}^2)$.

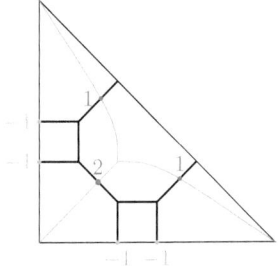
(f) The intersection $r \cdot (\varphi \cdot \mathbb{TP}^2)$. The sum of all weights is zero.

FIGURE 5. The intersection of a rational function with a Cartier divisor in $\mathbb{TP}^2$.

as r_{Q_1} is $-x$ and $-v_{Q_1/q_1} = (1,0)$. For similar reasons, the weights on q_2, q_3 and q_4 are also one. The point $q_5 = [(-\infty, 1, -1)]$ is the infinite face of $Q_5 = (-1, 2) + \mathbb{R}_{\geq 0}(1,1)$. However, r_{Q_5} is zero, so $w(q_5) = 0$ and, by symmetry, $w(q_6) = 0$.

Theorem 2.50. *Let φ and ψ be tropical Cartier divisors on C such that φ restricts to $\psi \cdot C$ and ψ restricts to $\varphi \cdot C$. Then $\psi \cdot \varphi \cdot C = \varphi \cdot \psi \cdot C$.*

PROOF. The general idea for this proof is as follows:
We have a maximal cell P, a facet Q_1 of P and a facet R of Q_1. By the diamond property of the face lattice of polytopes, there exists exactly one more facet Q_2 of P such that R is a facet of Q_2. Swapping Q_1 with Q_2 in the formulas computing the weight of R will then be paramount to switching between $\psi \cdot \varphi \cdot C$ and $\varphi \cdot \psi \cdot C$ since the relative data does not change, e.g. the normal vector v_{P/Q_1} is identical to $v_{Q_2/R}$.

We can assume without loss of generality that C has only components in one torus orbit and that this is the torus itself. There are four kinds of components in the products $\varphi \cdot \psi \cdot C$ and $\psi \cdot \varphi \cdot C$.

(1) Cells R which are facets of facets of cells of C. For these kind of cells the commutativity follows from [**AR09**, Prop. 6.7] as no infinite faces are involved.
(2) Cells P'' which are infinite faces of infinite faces of cells of C. The weight of such a cell can be computed as
$$w_{\psi\varphi}(P'') = w(P) \left[N(\sigma)_{P'} : N_P(\sigma)\right] \left[N(\tau)_{P''} : N_{P'}(\tau)\right] \varphi_P(v_{P/P'}) \psi_{P'}(v_{P'/P''}).$$
As $v_{P'/P''}$ is a vector from rec P there must be another infinite face $P^\star$ of P with $v_{P/P^\star} = v_{P'/P''}$ and $v_{P^\star/P''} = v_{P/P'}$. We also have
$$\left[N_P(\sigma) : N(\sigma)_{P'}\right] \left[N_{P'}(\tau) : N(\tau)_{P''}\right] = \left[N(\tau)_{P''} : N_P(\tau)\right].$$
Hence $w_{\varphi\psi}(P'') = w_{\psi\varphi}(P'')$.
(3) Cells Q' which are infinite faces of facets of cells of C or
(4) cells Q' which are facets of infinite faces of cells of C.

Every cell Q' can be obtained both as an infinite face Q' of a facet Q of a maximal cell P or as the facet Q' of an inifinite face P' of a maximal cell P.
Thus the weight of Q' is the sum of both of these constructions:
$$\begin{aligned} w_{\varphi\psi}(Q') &= w(Q) \left[N_Q(\sigma) : N(\sigma)_{Q'}\right] \varphi_Q(v_{Q/Q'}) \\ &+ \sum w(P') \varphi_{P'}(v_{P'/Q'}) \end{aligned}$$
with
$$w(Q) = \sum w(P) \psi_P(v_{P/Q})$$
and
$$w(P') = w(P) \left[N(\sigma)_{P'} : N_P(\sigma)\right] \psi_P(v_{P/P'}).$$
Hence we have
$$\begin{aligned} w_{\varphi\psi}(Q') &= \sum w(P) \psi_P(v_{P/Q}) \left[N_Q(\sigma) : N(\sigma)_{Q'}\right] \varphi_Q(v_{Q/Q'}) \\ &+ \sum w(P) \left[N(\sigma)_{P'} : N_P(\sigma)\right] \psi_P(v_{P/P'}) \varphi_{P'}(v_{P'/Q'}). \end{aligned}$$

Using the fact that
$$\varphi_{P'}([N(\sigma)_{P'}:N_P(\sigma)]\,v_{P'/Q'}) = \varphi_P([N(\sigma)_{Q'}:N_Q(\sigma)]\,v_{P/Q})$$
and $v_{P/P'} = v_{Q/Q'}$ we can rewrite this as
$$\begin{aligned} w_{\varphi\psi}(Q') &= \sum w(P)\psi_{P'}(v_{P'/Q'})\,[N_P(\sigma):N(\sigma)_{P'}]\,\varphi_P(v_{P/P'}) \\ &+ \sum w(P)\,[N(\sigma)_{Q'}:N_Q(\sigma)]\,\psi_Q(v_{Q/Q'})\varphi_P(v_{P/Q}) \\ &= w_{\psi\varphi}(Q'). \end{aligned}$$
□

Definition 2.51 (Push-Forward)**.** Let $f : \mathbf{X}_F(\mathbb{T})$ to $\mathbf{X}_G(\mathbb{T})$ be a morphism of complete tropical toric varieties. Let C be cycle on $\mathbf{X}_F$. We define a cycle $f_*(C)$ as follows: f maps every orbit $O(\sigma)$ of $\mathbf{X}_F(\mathbb{T})$ to an orbit $O(\tau)$ of $\mathbf{X}_G(\mathbb{T})$. We denote the corresponding linear map with f^σ. We have $C = \sum_{\sigma \in F} C_\sigma$ and send it to $f_*(C) = \sum_{\sigma \in F} f_*^\sigma(C_\sigma)$.

Definition 2.52 (Pull-Back)**.** Let $f : \mathbf{X}_F(\mathbb{T})$ to $\mathbf{X}_G(\mathbb{T})$ be a morphism of complete tropical toric varieties. Let $\varphi = (U_\alpha, r_\alpha)$ be a Cartier divisor on $\mathbf{X}_G(\mathbb{T})$. We define a Cartier divisor $f^*\varphi = (f^{-1}(U_\alpha), f^*r_\alpha)$ on $\mathbf{X}_F(\mathbb{T})$.

We have the following relation between push-forwards and pull-backs (extending [**AR09**, Prop. 4.8]).

Theorem 2.53 (Projection Formula)**.** *Let* $f : \mathbf{X}_F(\mathbb{T})$ *to* $\mathbf{X}_G(\mathbb{T})$ *be a morphism of complete tropical toric varieties. Let* $\varphi = (U_\alpha, r_\alpha)$ *be a Cartier divisor on* $\mathbf{X}_G(\mathbb{T})$.

Let E be a cycle on $\mathbf{X}_F(\mathbb{T})$. Then $\varphi \cdot f_*(E) = f_*(f^*\varphi \cdot E)$.

PROOF. We assume without loss of generality that E has only components in one torus orbit and that this is the main torus. The projection formula has been proven in [**AR09**, Prop 4.8, Prop 7.7] for all faces in the main torus. Hence we only need to compare the weights on infinite faces. Let σ be a cone of F. The image $f(\sigma)$ is contained in a cone τ of G. We denote the lattice of F with N and the lattice of G with K.

The weight of an infinite facet of E is $w(P') = w(P)\,[N(\sigma)_{P'}:N_P(\sigma)]\,f^*\varphi(v_{P/P'})$. The weight of the push-forward is
$$w_{f_*f^*\varphi E}(f(P')) = w(P)\,[K(\tau)_{f(P')}:f(N(\sigma)_{P'})]\,[N(\sigma)_{P'}:N_P(\sigma)]\,f^*\varphi(v_{P/P'}).$$

The weight of a cell $f(P)$ of the push-forward $f_*(E)$ is $w(P)\,\left[N'_{f(P)}:f(N_P)\right]$.
The weight of an infinite facet is
$$w_{\varphi f_* E}(f(P)') = w(P)\,\left[K_{f(P)}:f(N_P)\right]\,\left[K(\tau)_{f(P)'}:K_{f(P)}(\tau)\right]\,\varphi(v_{f(P)/f(P)'}).$$

If $v \in \operatorname{rec} P \cap \sigma$ then $f(v) \in \operatorname{rec} f(P) \cap f(\sigma)$, hence $f(P') = f(P)'$ and $f(v_{P/P'})$ and $v_{f(P)/f(P)'}$ are multiples of each other. Let us first compare $f(v_{P/P'})$ with $v_{f(P)/f(P)'}$. One is in the lattice $f(N_P \cap \operatorname{span} \sigma)$ while the other is in the lattice $K_{f(P)} \cap \operatorname{span} \tau$. Hence

$$\left[K_{f(P)} \cap \operatorname{span} \tau : f(N_P) \cap \operatorname{span} \sigma\right] v_{f(P)/f(P)'} = f(v_{P/P'}).$$

We therefore have to compare

$$\left[K(\tau)_{f(P')} : f(N(\sigma)_{P'})\right] \left[N(\sigma)_{P'} : N_P(\sigma)\right] \left[K_{f(P)} \cap \operatorname{span} \tau : f(N_P) \cap \operatorname{span} \sigma\right]$$

with

$$\left[K_{f(P)} : f(N_P)\right] \left[K(\tau)_{f(P)'} : K_{f(P)}(\tau)\right].$$

By a standard result of linear algebra we have

$$\left[K(\tau)_{f(P')} : f(N(\sigma)_{P'})\right] \left[K_{f(P)} \cap \operatorname{span} \tau : f(N_P \cap \operatorname{span} \sigma)\right] = \left[K_{f(P)} : f(N_P)\right].$$

Hence we only need to show

$$[N(\sigma)_{P'} : N_P(\sigma)] \overset{!}{=} \left[K(\tau)_{f(P)'} : K_{f(P)}(\tau)\right].$$

As f is injective on P', this is equivalent to

$$\left[f(N)(\tau)_{f(P)'} : f(N)_{f(P)}(\tau)\right] \overset{!}{=} \left[K(\tau)_{f(P)'} : K_{f(P)}(\tau)\right].$$

We can factor

$$\left[K(\tau)_{f(P)'} : f(N_P(\sigma))\right] = \left[K(\tau)_{f(P)'} : K_{f(P)}(\tau)\right] \left[K_{f(P)}(\tau) : f(N_P(\sigma))\right]$$
$$\left[K(\tau)_{f(P)'} : f(N_P)(\tau)\right] = \left[K(\tau)_{f(P)'} : f(N)(\tau)_{f(P)'}\right] \left[f(N)(\tau)_{f(P)'} : f(N_P)(\tau))\right]$$

and therefore show alternatively

$$\left[K(\tau)_{f(P)'} : f(N)(\tau)_{f(P)'}\right] \overset{!}{=} \left[K_{f(P)}(\tau) : f(N_P(\sigma))\right].$$

We factor again

$$\left[K(\tau)_{f(P)'} : f(N)(\tau)_{f(P)'}\right] = \left[K(\tau)_{f(P)'} : K_{f(P)}(\tau) + f(N)(\tau)_{f(P)'}\right]$$
$$\cdot \left[K_{f(P)}(\tau) + f(N)(\tau)_{f(P)'} : f(N)(\tau)_{f(P)'}\right]$$

and

$$\left[K_{f(P)}(\tau) : f(N_P(\sigma))\right] = \left[K_{f(P)}(\tau) : K_{f(P)}(\tau) \cap f(N)(\tau)_{f(P')}\right]$$
$$\cdot \left[K_{f(P)}(\tau) \cap f(N)(\tau)_{f(P')} : f(N_P(\sigma))\right].$$

We then see

$$\left[K_{f(P)}(\tau) + f(N)(\tau)_{f(P)'} : f(N)(\tau)_{f(P)'}\right] = \left[K_{f(P)}(\tau) : K_{f(P)}(\tau) \cap f(N)(\tau)_{f(P')}\right]$$

and
$$\left[K(\tau)_{f(P)'} : K_{f(P)}(\tau) + f(N)(\tau)_{f(P)'}\right] = \left[K_{f(P)}(\tau) \cap f(N)(\tau)_{f(P')} : f(N)_{f(P)}(\tau))\right]$$
which proves the theorem. □

4. Chow Groups

Definition 2.54 (Rational Equivalence). Let X be a tropical toric variety. We define a subgroup
$$R_k(X) = \mathrm{span}_{\mathbb{Z}} \{r \cdot C \mid r \in \mathrm{Rat}(X), C \in Z_{k+1}(X)\}$$
of $Z_k(X)$ of cycles that are generated by rational functions.

Following the treatment in [**AR09**], we define another subgroup
$$R'_k(X) = \mathrm{span}_{\mathbb{Z}} \{f_*(C) \mid f : Y \to X \text{ toric morphism}, C \in R_k(Y)\}$$
of $Z_k(X)$, obviously including $R_k(X)$. We then define the k-th Chow group of X as $A_k(X) := Z_k(X)/R'_k(X)$.

Remark 2.55. It is not a priori obvious whether $R_k(X)$ and $R'_k(X)$ are equal. The difference is that modding out by R'_k guarantees that a push-forward of elements equivalent to zero is again equivalent to zero. See [**AR09**, Remark 8.6] for an example why this is necessary. We will not consider the group $R_k(X)$ again; rational equivalence will always mean equivalence with respect to the larger group $R'_k(X)$.

Remark 2.56. One can also consider the group $Z_k(Y)$ of k-dimensional subcycles of a cycle Y of a tropical toric variety X. In oder to form a Chow group $A_k(Y)$, we should mod out by all push-forwards of divisors of rational functions into Y. For this, we need maps between cycles, and these should probably be locally linear maps as in [**AR09**, Def. 7.1], i.e. locally morphisms of toric varieties.

We now show that the intersection product is well-defined up to rational equivalence.

Theorem 2.57.
(1) Let φ be a rational function on a k-cycle C. Then $\varphi \cdot C \sim 0$.
(2) Let C be a k-cycle equivalent to zero and φ a Cartier divisor on C. Then $\varphi \cdot C$ is equivalent to zero.
(3) Let $f : C \to D$ be a morphism and C equivalent to zero. Then $f_*(C)$ is equivalent to zero.
(4) Let $f : C \to D$ be a surjective morphism and φ a rational function on D. Then $f^*\varphi$ is a rational function on C.

PROOF.

(1) This follows from the definition of rational equivalence.
(2) We have $C = f_*(\psi \cdot E)$ with ψ a rational function on E. Then $\varphi \cdot C = f_*(f^*\varphi \cdot \psi \cdot E) = f_*(\psi \cdot f^*\varphi \cdot E)$ which is equivalent to zero by definition.
(3) We have $C = g_*(\psi \cdot E)$ and $f_*(C) = f_*(g_*(\psi \cdot E)) = (f \circ g)_*(\psi \cdot E)$.
(4) $f^*\varphi : D \to \mathbb{R}, x \mapsto \varphi(f(x))$ is a piece-wise affine linear function.

$\square$

We will show to main results about rational equivalence on a tropical toric variety $\mathbf{X}_F(\mathbb{T})$:

(1) Every cycle in the torus T is equivalent to a cycle in the boundary, it is a formal sum of orbit closures.
(2) Every cycle in the boundary is equivalent to a tropical fan in T, this fan can be chosen to be a subfan of F.

These results correspond to the classical duality between the groups $A_k(\mathbf{X}_F(\mathbb{C}))$ of torus-invariant subvarieties and $\mathrm{MW}_k(F)$ of Minkowski weights.

Lemma 2.58. *Let X be a smooth complete tropical toric variety with torus T and C a tropical complex in T of codimension at least one. Then C is rationally equivalent to a cycle that has no components in T.*

PROOF. Let C be a tropical complex of codimension at least one in the torus T of the tropical variety X.

We choose a vector $a \in N$. We construct a new cycle $\tilde{C} = \{(x + \lambda a, \lambda) \mid x \in C, \lambda \in \mathbb{R}\}$ in the torus $T \times \mathbb{R}$ of the toric variety $Y = X \times \mathbb{TP}^1$.

We consider the toric morphism $\mathrm{pr}_1 : X \times \mathbb{TP}^1 \to X$ that forgets the second factor. Let ρ, ρ' be the rays of the second factor.

We use $\varphi = \max(0, x_\rho)$ as a rational function on Y. We then find $\varphi \cdot Y = (-D_\rho + [\varphi]) \cdot Y$. Hence the two cycles $D_\rho \cdot \tilde{C}$ and $[\varphi] \cdot \tilde{C}$ are rationally equivalent. $[\varphi] \cdot \tilde{C}$ is equal to $C \times \{0\}$.

Let us look at $D_\rho \cdot \tilde{C} = \sum_{\sigma \in F} O(\sigma + \rho) \cdot \tilde{C}$. We want to show $O(0 + \rho) \cdot \tilde{C} = 0$ for suitable choices of a.

Let P be a maximall cell of C. This leads to two maximal cells $(P, 0) + \mathbb{R}_{\geq 0}(a, 1)$, $(P, 0) + \mathbb{R}_{\leq 0}(a, 1)$ of $\tilde{C}$. We need to check whether $\tilde{P} = (P, 0) + \mathbb{R}_{\geq 0}(a, 1)$ gives rise to a cell P' of $O(0 + \rho) \cdot \tilde{C}$. That means we need to compute the dimension of $(P, 0) + \mathbb{R}_{\geq 0}(a, 1)/\mathrm{span}(0, 1) \cong P + \mathbb{R}_{\geq 0}a$. This dimension is $\dim P$ if $a \in \mathrm{span}(\mathrm{rec}\, P)$ and $1 + \dim P = \dim \tilde{P}$ otherwise.

Hence, if we take a outside of the finitely many linear subspaces spanned by the recession cones of the maximal cells of C, then $O(0 + \rho) \cdot \tilde{C}$ is empty.

So let us assume there is a polyhedron P' in $O(0 + \rho) \cdot \tilde{C}$. That means there is a polyhedron $\tilde{P}$ that is a maximal cell of $\tilde{C}$ with $P' = \tilde{P}/\operatorname{span}(\sigma)$. This in turn comes from a maximal cell P of C with $\tilde{P} = (P + \mathbb{R}_{\geq 0}a, \mathbb{R}_{\geq 0})$.

This means $C = \operatorname{pr}_1([\varphi]\cdot\tilde{C})$ is equivalent to $\operatorname{pr}_1(D_\rho\cdot\tilde{C})$, which is a cycle that has no components in the orbit $T = O(0)$ of X. □

Theorem 2.59. *In a smooth complete tropical toric variety every cycle is equivalent to a formal sum of orbit closures.*

PROOF. We start with an arbitrary cycle A_0. If A_0 is not a sum of orbit closures, then it contains a polyhedral complex C_0 of codimension at least one in some orbit $O(\sigma)$. We then apply Lemma 2.58 to C_0 in $V(\sigma)$ and arrive at a cycle A_1. We repeat until Lemma 2.58 can no longer be applied. Thus the resulting cycle is a sum of orbit closures. □

Corollary 2.60. *Let X be a compact smooth tropical toric variety. Then $A_k(X)$ is generated by the k-dimensional orbit closures of X.*

Lemma 2.61. *Let X be a complete smooth tropical toric variety. Let M be a Minkowski weight and $[M]$ the class of M under rational equivalence. Then $[M] = 0$ if and only if $M = 0$.*

PROOF. This is the result of [**AR08**, Lemma 6]. Note that [**AR08**] uses a weaker notion of rational equivalence. The result applies to our situation since only one torus orbit occurs. One could also modify the proof to our definition of rational equivalence. □

Definition 2.62 (Support Functions on Fans). Let F be a complete unimodular fan with lattice N and let $T = N \otimes \mathbb{R}$. Let $h : T \to \mathbb{R}$ be a tropical rational function that is linear on every cone of F. Such an h is called a support function on F.

Lemma 2.63. *A support function h for an n-dimensional unimodular fan F is uniquely characterized by a collection $(a_\rho)_{\rho \in F^{(1)}}$ of integers with $h(-v_\rho) = a_\rho$ or a collection $(m_\sigma)_{\sigma \in F^{(n)}}$ of elements from $N^\vee$ with $h(v) = m_\sigma v$ for all $v \in \sigma$.*

PROOF. The statement is proven in [**Ful93**, Section 3.4]. This also follows from Lemma 2.41 and Lemma 2.42. □

Corollary 2.64. *For every sum $\sum a_\rho D_\rho$ of boundary divisors there is a Cartier divisor φ_a with $\varphi_a \cdot \mathbf{X}_F(\mathbb{T}) = \sum a_\rho D_\rho$.*

Definition 2.65. Let F be a complete unimodular fan and h a support function on F. We construct a Cartier divisor φ_h via the covering of X with (U_σ, m_σ) with the notation of the previous lemma. Alternatively, we can set $\varphi_h = \sum a_\rho D_\rho$.

Lemma 2.66. *Let F be a complete smooth fan and h a support function on F. Then $h \cdot \mathbf{X}_F(\mathbb{T}) = M + \varphi_h \cdot \mathbf{X}_F(\mathbb{T})$ where M is a Minkowski weight.*

PROOF. The function h is linear on each maximal cone of F, hence $[h] \cdot T$ is a subfan of F. Since $\mathrm{ord}_\rho(h) = h(-v_\rho) = a_\rho$ we have the claimed result. $\square$

Corollary 2.67. *Let F be a complete unimodular fan and $\mathbf{X}_F(\mathbb{T})$ the corresponding tropical variety.*

(1) *The group $A_k(\mathbf{X}_F(\mathbb{T}))$ is generated by the classes of k-dimensional Minkowski weights.*
(2) *The group $A_k(\mathbf{X}_F(\mathbb{T}))$ is isomorphic to the group $\mathrm{MW}_k(F)$ of k-dimensional Minkowski weights.*
(3) *The group $A_k(\mathbf{X}_F(\mathbb{T}))$ is isomorphic to the classical Chow group $A_k(\mathbf{X}_F(\mathbb{C}))$.*

PROOF. Let n be the dimension of $\mathbf{X}_F(\mathbb{T})$. We consider an inclusion map $i : \mathrm{MW}_k(F) \to A_k(X)$. This map is injective via Lemma 2.61. The map $\mathrm{MW}_{n-1}(F) \to A_{n-1}(\mathbf{X}_F(\mathbb{T}))$ is surjective via Lemma 2.63 and Lemma 2.66. Surjectivity for higher codimension follows from the fact that every boundary divisor of codimension k is the intersection product of k boundary divisors of codimension one. The group of Minkowski weights is isomorphic to $\hom(A_k(Y), \mathbb{Z})$ by [**FS97**, Prop. 1.4]. $\square$

Remark 2.68. Since every cycle is equivalent to a sum of products of Cartier divisors, we can intersect arbitrary cycle classes and get intersection products

$$A_{n-p}(\mathbf{X}_F(\mathbb{T})) \times A_{n-q}(\mathbf{X}_F(\mathbb{T})) \to A_{n-p-q}(\mathbf{X}_F(\mathbb{T})).$$

Furthermore, the diagonal $\Delta \subseteq \mathbf{X}_F(\mathbb{T}) \times \mathbf{X}_F(\mathbb{T})$ is equivalent to a sum of products of Cartier divisors, we can define the intersection product as

$$[C] \cdot [D] := \mathrm{pr}_*([\Delta] \cdot X \times Y)$$

(the calculus of this is worked out in the proof of [**AR08**, Theorem 9.10] and preceeding lemmata).

This construction might be more economical in practice as one does not have to rewrite arbitrary cycles as Cartier divisors. Another advantage is that this allows us to define an intersection product for cycle classes in every ambient tropical polyhedral complex E such that the class of the diagonal $\Delta \subset E \times E$ can be expressed as a sum of products of Cartier divisors (for example those polyhedral complexes that are locally isomorphic to tropical linear spaces satisfy this condition).

CHAPTER 3

Tropicalization

In this chapter we will relate the tropical intersection theory developed in Chapter 2 to the intersection theory of complex and non-Archimedean toric varieties. A central object will be the tropicalization of an algebraic variety.

Definition 3.1 (Tropicalization). Let $\mathbb{K}$ be an algebraically closed field with a non-trivial non-Archimedean valuation val : $\mathbb{K}^\times \to \mathbb{R}$, that is we have

(1) $\operatorname{val}(a \cdot b) = \operatorname{val}(a) + \operatorname{val}(b)$.
(2) $\operatorname{val}(a + b) \leq \max(\operatorname{val}(a), \operatorname{val}(b))$.
(3) There is a $t \in \mathbb{K}$ such that $\operatorname{val}(t) = 1$.

This map extends to a map of semigroups val : $\mathbb{K} \to \mathbb{T}$ by setting $\operatorname{val}(0) = -\infty$. Applying val component-wise, we get a map val : $\mathbb{K}^n \to \mathbb{T}^n$.

Let $X \subseteq \mathbb{K}^n$ be an affine algebraic variety. Then the topological closure of the image of X under valuation

$$\operatorname{trop} X := \overline{\operatorname{val}(X)}$$

is called the tropicalization of X.

Example 3.2. When thinking of a non-Archimedean field in tropical geometry, one should think of the field $\mathbb{C}\{\{t\}\}$ of complex Puiseux series. It is defined as the limit

$$\mathbb{C}\{\{t\}\} = \bigcup_{n \geq 1} \mathbb{C}((t^{\frac{1}{n}}))$$

where $\mathbb{C}((t))$ is the field of formal Laurent series, the quotient field of the ring of formal power series.

The valuation of an element $f = \sum_{k=0}^{\infty} a_k t^{\frac{k}{n}}$ is

$$\operatorname{val} f := -\operatorname{ord} f := -\min\left\{\frac{k}{n} \mid a_k \neq 0\right\}.$$

We have $\operatorname{val}(\mathbb{C}\{\{t\}\}) = \mathbb{Q}$.

Let X be a variety in the torus $(\mathbb{K}^\times)^n \subseteq \mathbf{X}_F(\mathbb{K})$. We want to study the closure of the tropicalization trop $X \subseteq \mathbb{R}^n$ inside $\mathbf{X}_F(\mathbb{T})$. The tropicalization itself is a closure of the valuation inside $\mathbb{R}^n$.

We can simplify the situation by choosing a field $\mathbb{K}$ that has a surjective valuation. In this case $\mathrm{trop}\, X$ equals $\mathrm{val}\, X$ and we can omit the topological closure inside $\mathbb{R}^n$.

Definition 3.3. The set
$$\mathbb{C}\{\{t^\mathbb{R}\}\} := \left\{ \sum_{\alpha \in A} a_\alpha t^\alpha \mid A \subseteq \mathbb{R} \text{ well-ordered}, a_\alpha \in \mathbb{C} \right\}$$
is an algebraically closed field of characteristic zero with surjective valuation
$$\mathrm{val} : \mathbb{C}\{\{t^\mathbb{R}\}\}^\times \to \mathbb{R}, \sum_{\alpha \in A} a_\alpha t^\alpha \mapsto \min\{\alpha \mid a_\alpha \neq 0\}.$$
It is called the field of transfinite Puiseux series and complete with respect to this valuation.

Generalizations of this field are used in [**Pay09b**], we refer to the references mentioned there for proofs that $\mathbb{C}\{\{t^\mathbb{R}\}\}$ has well-defined field operations and is algebraically closed. Another field with a surjective valuation to $\mathbb{R}$ is constructed in [**Mar07**].

We have the following result of [**AN09**], where the characteristic pair of a non-Archimedean field $\mathbb{K}$ is the pair of characteristics (char $\mathbb{K}$, char $\Bbbk$) where $\Bbbk$ is the residue field
$$\Bbbk = \{x \in \mathbb{K} \mid \mathrm{val}(x) \leq 0\}/\{x \in \mathbb{K} \mid \mathrm{val}(x) < 0\}$$
of the valuation ring modulo its unique maximal ideal.

Theorem 3.4. [**AN09**, Theorem 4.4] *The set of tropical varieties definable over an algebraically closed valued field only depends on the characteristic pair of the valued field and on the image of the valuation. If one considers valuations that are surjective on $\mathbb{R}$, the set of tropical varieties only depends on the characteristic pair.*

Hence we can use our intuition for the field $\mathbb{C}\{\{t\}\}$, where up to scaling every point $x \in \mathbb{C}\{\{t\}\}$ is just a formal power series while using the field $\mathbb{C}\{\{t^\mathbb{R}\}\}$ for technical reasons.

Theorem 3.5. [**BJS**$^+$**07**, Theorem 1.2] *Let $X \subseteq (\mathbb{K}^\times)^n$ be an irreducible variety of dimension d. Then $\mathrm{trop}\, X \subseteq \mathbb{R}^n$ is a connected polyhedral complex of pure dimension d.*

The most important property of tropicalization for our purposes is that it produces tropical complexes as in Definition 2.2.

Theorem 3.6. *Let $X \subseteq (\mathbb{K}^\times)^n$ be an irreducible variety of dimension d. Then $\mathrm{trop}\, X \subseteq \mathbb{R}^n$ is a balanced polyhedral complex.*

PROOF. Proofs are in [**Spe05**, Theorem 2.1.5] or [**Kat09b**, Theorem 8.14] □

There is an interesting special case of this theorem, in the case that $\mathrm{trop}\, X$ is a fan.

Lemma 3.7. *Let $X = \mathbf{X}_F(\mathbb{K})$ be a smooth complete toric variety with torus $T \cong (\mathbb{K}^\times)^n$. Let C be an irreducible subvariety of T. Assume that there is a subfan G of F with $|\operatorname{trop} C| = |G|$. Then $\operatorname{MW}([C]) = \operatorname{trop} C$ where $\operatorname{MW}([C])$ denotes the Minkowski weight corresponding to the map $A_{n-\dim C}(X) \to \mathbb{Z}$ defined via $[Y] \mapsto [C] \cdot [Y] \in A_0(X) \cong \mathbb{Z}$.*

PROOF. This is proven in [**Kat09b**, Prop. 9.4] (using a different sign convention). □

Definition 3.8. Let $\mathbb{K}$ be a field with non-Archimedean valuation and let $f = \sum_{j \in J} a_j x^j \in \mathbb{K}[x_1^{\pm 1}, \ldots, x_n^{\pm 1}]$ be a Laurent polynomial. We define the set $\mathcal{T}(f) \subseteq \mathbb{R}^n$ via the following condition:

A point $w \in \mathbb{R}^n$ lies in $\mathcal{T}(f)$ if and only if the maximum in $\max\{\operatorname{val}(a_j) + j \cdot w \mid j \in J\}$ is achieved at least twice. Note that $\mathcal{T}(f)$ depends only on the tropicalization of f.

If $I \subseteq K[x_1, \ldots, x_n]$ is an ideal, we set

$$\mathcal{T}(I) = \bigcap_{f \in I} \mathcal{T}(f)$$

A finite generating set $f_1, \ldots, f_k$ of I is called a tropical basis if $\mathcal{T}(I) = \mathcal{T}(f_1) \cap \ldots \cap \mathcal{T}(f_k)$.

Theorem 3.9. [**BJS$^+$07**, Theorem 2.9] *Every ideal $I \subseteq K[x_1, \ldots, x_n]$ has a tropical basis.*

Theorem 3.10. [**SS04**, Theorem 2.1] *Let $X \subseteq (\mathbb{K}^\times)^n$ be an irreducible variety of dimension d. The following sets are equal*

(1) $\operatorname{trop} X$
(2) $\{w \in \mathbb{R}^n \mid \operatorname{in}_w(I(X))$ *contains no monomial*$\}$
(3) $\mathcal{T}(I)$.

One should also mention that there is a continuous surjection from the Berkovich space associated to X to $\operatorname{trop} X$ [**Spe05**, Prop. 2.1.5.].

We will now extend these notion from tori to toric varieties (this was called extended tropicalization in [**Pay09a**]).

Definition 3.11. Let F be a rational fan. We can extend the valuation $\operatorname{val} : \mathbb{K} \to \mathbb{T}$ to a map $\operatorname{val} : \mathbf{X}_F(\mathbb{K}) \to \mathbf{X}_F(\mathbb{T})$ in the following way:

For each affine open set U_σ we get a semigroup homomorphism

$$\operatorname{val} : \hom(S_\sigma, \mathbb{K}) \to \hom(S_\sigma, \mathbb{T}), \, f \mapsto \operatorname{val} \circ f.$$

We can glue these maps together to get a map $\operatorname{val} : \mathbf{X}_F(\mathbb{K}) \to \mathbf{X}_F(\mathbb{T})$. Note that this map restricts to a map $\operatorname{val} : O_\mathbb{K}(\sigma) \to O_\mathbb{T}(\sigma)$ on each torus orbit. We set

$$\operatorname{trop} Y := \overline{\operatorname{val}(Y)}$$

for any subvariety $Y \subseteq \mathbf{X}_F(\mathbb{K})$.

Remark 3.12. If σ is a unimodular cone of dimension k of an n-dimensional fan F, then any choice of a lattice basis with rays of σ leads to the usual component-wise valuation

$$U_\mathbb{K}(\sigma) \cong \mathbb{K}^k \times (\mathbb{K}^\times)^{n-k} \to \mathbb{T}^k \times \mathbb{R}^{n-k} \cong U_\mathbb{T}(\sigma).$$

Definition 3.13. Let $X = \mathbf{X}_F(\mathbb{K})$ be a simplicial toric variety and $I \subseteq \mathbb{K}[F^{(1)}]$ be a homogeneous ideal. The set $V_X(I) = \{[p] \in X \mid f(p) = 0 \text{ for all } f \in I\}$ is called the zero-set of I. If $f = \sum a_J x^J$ is a polynomial in I, we set

$$\mathcal{T}_X(f) := \{[p] \in \mathbf{X}_F(\mathbb{T}) \mid \text{The maximum in } \max(\text{val } a_J + J \cdot x) \text{ is achieved twice}\}.$$

We set $\mathcal{T}_X(I) = \bigcup_{f \in I} \mathcal{T}_X(f)$.

Theorem 3.14. [Cox95, Prop. 2.4] $V_X(I) = \emptyset$ if and only if I contains a power of the irrelevant ideal. Let $X = \mathbf{X}_F(\mathbb{K})$ be a simplicial toric variety. There is a one-to-one correspondence between subvarieties of X and radical homogeneous ideals of $\mathbb{K}[F^{(1)}]$ contained in the irrelevant ideal.

Corollary 3.15. Let $X = \mathbf{X}_F(\mathbb{K})$ be a simplicial toric variety and $I \subseteq \mathbb{K}[F^{(1)}]$ be a homogeneous ideal. Then $\text{trop } V_X(I) = \mathcal{T}_X(I)$. For each torus orbit $O(\sigma)$ we have

$$\text{trop } V_X(I) \cap O_\mathbb{K}(\sigma) = \{w \in O_\mathbb{T}(\sigma) \mid \text{in}_w(I_\sigma) \text{ contains no monomial}\}.$$

Lemma 3.16. Let F be a smooth complete fan and $\sigma \in F$ a cone. Let $O_\mathbb{K}(\tau)$ be the torus orbit of $\mathbf{X}_F(\mathbb{K})$ and $O_\mathbb{T}(\tau)$ the torus orbit of $\mathbf{X}_F(\mathbb{T})$ corresponding to the cone τ of F.

Then $\text{trop}(O_\mathbb{K}(\sigma)) = O_\mathbb{T}(\sigma)$. In particular, for Weil divisors corresponding to rays ρ: $\text{trop}(D_\rho) = D_\rho$ (or, more precisely, $\text{trop}(V_\mathbb{K}(\rho)) = V_\mathbb{T}(\rho)$).

Corollary 3.17. trop induces isomorphisms $\text{trop}: A_k(\mathbf{X}_F(\mathbb{K})) \to A_k(\mathbf{X}_F(\mathbb{T}))$ on the classes of boundary divisors.

Theorem 3.18. Let F be an n-dimensional complete smooth fan and $\mathbf{X}_F(\mathbb{K})$ the corresponding toric variety. Let f be a homogeneous polynomial in the Cox ring of $\mathbf{X}_F(\mathbb{K})$.
Then

$$[Z(f)] = [\text{trop } Z(f)] = \text{trop } f \cdot [\mathbf{X}_F(\mathbb{T})]$$

where we identify $A_{n-1}(\mathbf{X}_F(\mathbb{K}))$ and $A_{n-1}(\mathbf{X}_F(\mathbb{T}))$.

PROOF. We have $\deg f = \deg \text{trop } f$ and $[Z(f)] = \deg f$. □

Corollary 3.19. Let F be a complete smooth fan and $\mathbf{X}_F(\mathbb{K})$ the corresponding non-Archimedean toric variety. Let C be the k-cycle of a tropically transverse complete intersection in $\mathbf{X}_F(\mathbb{K})$.

Then $[C] = [\text{trop } C]$.

Theorem 3.20. [**Kat09b**, Theorem 8.8] *Let $F \subseteq \mathbb{R}^n$ be a complete smooth fan and $\mathbf{X}_F(\mathbb{K})$ the corresponding non-Archimedean toric variety. Let C be an irreducible subvariety of dimension k and D an irreducible subvariety of dimension $n - k$ that intersect transversally. Assume further that $\operatorname{trop} C$ and $\operatorname{trop} D$ intersect transversally.*
Then $[C] \cdot [D] = [\operatorname{trop} C] \cdot [\operatorname{trop} D]$.

Theorem 3.21. *Let F be a complete smooth fan and $\mathbf{X}_F(\mathbb{K})$ the corresponding non-Archimedean toric variety. Let C be an irreducible subvariety.*
Then $[C] = [\operatorname{trop} C]$.

PROOF. Let $n = \dim \mathbf{X}_F(\mathbb{K})$ and $k = \dim C$.
We know that $[C]$ is completely determined by the products $[C] \cdot [D_i]$ where the D_i generate $A_{n-k}(\mathbf{X}_F(\mathbb{K}))$. The same is true for $\operatorname{trop} C$ and $A_{n-k}(\mathbf{X}_F(\mathbb{T}))$.

We know that there exists a finite set of $H_1, \ldots, H_s$ of hypersurfaces such that the $[H_j]$ generate $A_{n-1}(\mathbf{X}_F(\mathbb{K}))$ and they intersect tropically transverse with each other and with C.

For every complete intersection $H_I = \bigcap_{i \in I} H_i$ of $n - k$ of the H_j we know: $\deg H_I = \deg \operatorname{trop} H_I$ and $\deg C \cap H_I = \deg \operatorname{trop} C \cdot \operatorname{trop} H_I$. This means $\deg C = \deg \operatorname{trop} C$. □

The connections between the closure of C in toric varieties and the fan $\operatorname{trop} C$ has also been used to produce compactifications of C that have a desirable combinatorial structure (e.g. $\overline{C} \backslash C$ is a sum of divisors and those divisors have normal crossings).

Let $T \cong (\mathbb{K}^\times)^d$ be an algebraic torus and $X \subseteq T$ a subvariety.

We want to find a compactification $\overline{X} \supseteq X$ that corresponds to a toric variety $\mathbf{X}_F(\mathbb{K}) \supseteq T$.

The idea in this section is to use the fan $\operatorname{trop} X$ to compactify X by including T in the toric variety $\mathbf{X}_{\operatorname{trop} X}(\mathbb{K})$. Usually, one only knows that there is a fan F with $|F| = \operatorname{trop} X$, but there is no canonical fan structure on the set $\operatorname{trop} X$. Hence for every choice of a fan structure F on $\operatorname{trop} X$ there is a corresponding tropical compactification of X.

This is the idea of [**Tev07**].

Of course, this construction also yields a compactification $\mathbf{X}_{\operatorname{trop} X}(\mathbb{T}) \supseteq \operatorname{trop} X$, which we will call the tropical compactification of $\operatorname{trop} X$.

Definition 3.22.

- Let $V \cong \mathbb{R}^n$ be a vector space and $X \subsetneq V$ a balanced fan with $|X| \subsetneq V$. Let F be a smooth rational fan of with $|F| = |X|$.
  Then the closure of X in $\mathbf{X}_F(\mathbb{T})$ is called a tropical compactification of X and denoted $\overline{X}_{\text{vc}} := \overline{X}_{\text{vc}}^F$.

47

- Let $\Bbbk$ be an algebraically closed field, T a torus over $\Bbbk$ and X a connected closed subvariety of T. Let $\mathbb{K}$ be an algebraically closed field containing $\Bbbk$ with surjective valuation val : $\mathbb{K} \to \mathbb{T}$ that is trivial on $\Bbbk$. Let $X(\mathbb{K})$ and $T(\mathbb{K})$ be the $\mathbb{K}$-valued points of X and T. Then we can consider the balanced fan trop $X(\mathbb{K}) \subseteq$ trop $T(\mathbb{K}) \cong \mathbb{R}^{\dim T}$. Let F be a smooth rational fan with $|F| = $ trop $X(\mathbb{K})$.

  Then the closure of X in $\mathbf{X}_F(\Bbbk)$ is a called a tropical compactification of X and denoted $\overline{X}_{\mathrm{vc}} := \overline{X}^F_{\mathrm{vc}}$.

Remark 3.23. Let $\Bbbk = \mathbb{C}$. The fan F is not a complete fan so the toric varieties $\mathbf{X}_F(\mathbb{C})$ and $\mathbf{X}_F(\mathbb{T})$ are not compact. We will show that $\overline{X}_{\mathrm{vc}}$ is compact.

Lemma 3.24. *Let F be a rational fan structure on the balanced polyhedral complex X in a vector space V. Then $\overline{X}^F_{\mathrm{vc}}$ is compact.*

PROOF. Let F' be a smooth rational fan refining F. There is a continuous map $\mathbf{X}_{F'}$ to $\mathbf{X}_F$ that we can restrict to $\overline{X}^{F'}_{\mathrm{vc}}$ and $\overline{X}^F_{\mathrm{vc}}$. Therefore $\overline{X}^F_{\mathrm{vc}}$ is compact if $\overline{X}^{F'}_{\mathrm{vc}}$ is compact. Hence we can assume without loss of generality that F is smooth.

Let $k = \dim F$. We have $|F| = \bigcup_{\sigma \in F^{(k)}} |\sigma|$. Hence we have to show that each closure $\overline{\sigma}$ is compact in $\mathbf{X}_F(\mathbb{T})$ for $\sigma \in F^{(k)}$. Let σ be any maximal cone of F. We look at $\overline{\sigma}$ in the chart U_σ. σ is generated by rays $r_1, \ldots, r_k$. Each ray r_i is represented by $-e_i$ in the chart $U_\sigma \cong \mathbb{T}^k$. Hence $\overline{\sigma} = \{x \in \mathbb{T}^k \mid -\infty \leq x_i \leq 0\} \approx [0,1]^k$ is compact. □

Remark 3.25. The classical analogue is [**Tev07**, Prop. 2.3]: the closure $\overline{X}$ of the subvariety X in $\mathbf{X}_F(\mathbb{K})$ is proper if and only if $|\mathrm{trop}\, X| \subseteq |F|$.

Theorem 3.26. [**Tev07**, Theorem 1.2] *Any subvariety X of a torus has a tropical compactification $\overline{X}$ in a smooth projective toric variety such that the boundary $\overline{X} \setminus X$ is divisorial and intersections of these divisors have the expected codimension.*

CHAPTER 4

Parameter Spaces of Lines in $\mathbb{TP}^n$

We want to look at the set of all tropical lines in $\mathbb{TP}^2$. In this case a tropical line is characterized by the following equivalent conditions

(1) it is the tropicalization of a one-dimensional linear space in $\mathbb{KP}^2$,
(2) it is the zero-set of a homogeneous tropical polynomial of degree one,
(3) it is a balanced polyhedral complex with non-negative weights that is rationally equivalent to the one-skeleton of $\mathbb{TP}^2$.

The most common case is the generic tropical line, which corresponds to

(1) the tropicalization of a one-dimensional linear space in $\mathbb{KP}^2$ that is generated by a vector in $(\mathbb{K}^\times)^2$,
(2) the zero-set of a homogeneous tropical polynomial of degree one where all coefficients are present and finite,
(3) a balanced polyhedral complex that is a translate of $V(x_0 \oplus x_1 \oplus x_2)$.

We can use $\mathbb{TP}^2$ as a parameter space for tropical lines by sending the point $[(a_0, a_1, a_2)]$ to the space $V(a_0 \odot x_0 \oplus a_1 \odot x_1 \oplus a_2 \odot x_2) \subseteq \mathbb{TP}^2$.

The situation is more complicated in higher dimensions, i.e. lines in $\mathbb{TP}^n$ or even k-dimensional tropical linear spaces in $\mathbb{TP}^n$. The situation is well-understood for linear spaces over a field, they are parametrized by the Grassmannian variety. We will study its tropicalization.

Definition 4.1.

- Let n be a natural number. We set $[n] := \{1, \ldots, n\}$.
- Let S be a finite set and k a natural number. Then $\binom{S}{k}$ is the collection of all subsets of S of size k. Hence $\#\binom{S}{k}$ is equal to the binomial coefficient $\binom{\#S}{k}$.

Definition 4.2 (Grassmannian)**.** Let $\mathbb{K}$ be a field. Let S be a finite set.

We consider the r-fold exterior product $\bigwedge^r \mathbb{K}^S \cong \mathbb{K}^{\binom{S}{r}}$.

The Grassmannian $\mathbf{G}_{r,S}(\mathbb{K})$ is the subset of $\mathbb{P}(\bigwedge^r \mathbb{K}^S)$ that consists of all decomposable vectors:

$$\mathbf{G}_{r,S} := \mathbf{G}_{r,S}(\mathbb{K}) := \left\{ p \in \mathbb{P}(\bigwedge^r \mathbb{K}^S) \mid \exists v_1, \ldots, v_r \in \mathbb{K}^S : p = v_1 \wedge \ldots \wedge v_r \right\}$$

If n is a natural number, then we set $\mathbf{G}_{r,n} := \mathbf{G}_{r,[n]}$

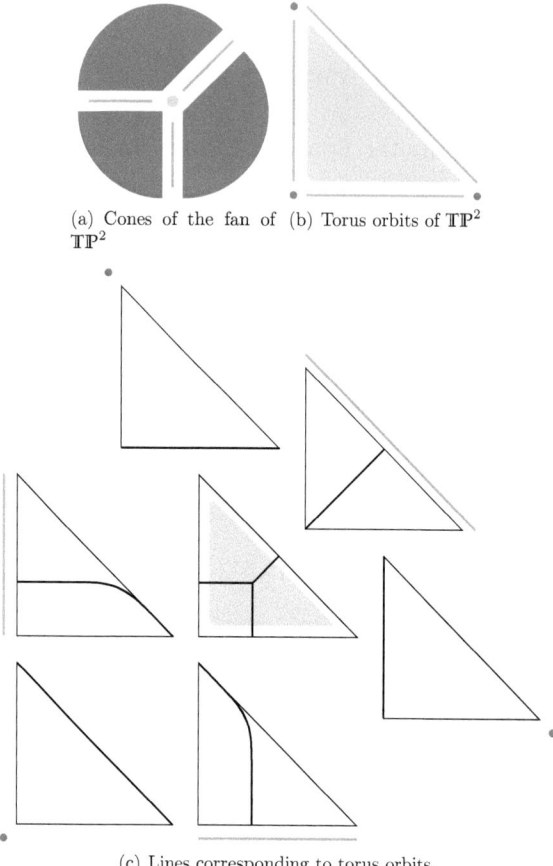

(a) Cones of the fan of $\mathbb{TP}^2$

(b) Torus orbits of $\mathbb{TP}^2$

(c) Lines corresponding to torus orbits.

FIGURE 6. Lines in $\mathbb{TP}^2$ correspond to torus orbits, translations of a line lie in the same orbit.

Remark 4.3. The Grassmannian is an algebraic variety that is cut out by the ideal of Plücker relations (see for example [**Har92**, Lecture 6], [**GKZ94**, Chapter 3] or [**Stu93**, Chapter 3]). We will only need the following case: Fix a total order $<$ on S. Let $\{x_I\}_{I \in \binom{S}{r}}$ be the homogeneous coordinates of $\mathbb{P}(\bigwedge^r \mathbb{K}^S)$. If $r = 2$ then the Plücker relations are generated by the polynomials

$$x_{ij}x_{kl} - x_{ik}x_{jl} + x_{il}x_{jk}$$

for all $i < j < k < l$.

Remark 4.4. Points in the Grassmannian $\mathbf{G}_{r,S}$ are in one-to-one correspondence with $(r-1)$-dimensional linear subspaces of $\mathbb{P}(\mathbb{K}^S)$:

- An $(r-1)$-dimensional plane in $\mathbb{P}(\mathbb{K}^S)$ corresponds to an r-dimensional linear subspace $L = \mathrm{span}(b_1, \ldots, b_r)$ in $\mathbb{K}^S$. To this space we associate the point $p = b_1 \wedge \ldots \wedge b_r$ (the class in $\mathbb{P}(\bigwedge^r \mathbb{K}^S)$ does not depend on the choice of a basis).
- To a point p in the Grassmannian we associate the linear space
$$L(p) := \left\{ v \in \mathbb{K}^S \mid v \wedge p = 0 \in \bigwedge\nolimits^{r+1} \mathbb{K}^S \right\}.$$

For $r = 2$ and $S = [n]$ we can describe this as
$$L(p) = \left\{ z \in \mathbb{P}^{n-1} \mid p_{ij}z_k - p_{ik}z_j + p_{jk}z_i = 0 \text{ for all } 1 \le i < j < k \le n \right\}$$
where $z_1, \ldots, z_n$ are homogeneous coordinates on $\mathbb{P}(\mathbb{K}^n)$.

Definition 4.5. The familiy $F\mathbf{G}_{r,S} := \{(p,z) \mid z \in L(p)\} \subseteq \mathbf{G}_{r,S} \times \mathbb{P}(S)$ is called the tautological bundle of the Grassmannian. It is an $(r-1)$-dimensional projective bundle over $\mathbf{G}_{r,S}$.

The Grassmannian $\mathbf{G}_{r,S}$ has an open dense subset $\mathbf{G}_{r,S}^\circ = \{x \mid x_I \ne 0 \text{ for all } I \subseteq S, |I| = r\}$. We will now look at its complement $\partial \mathbf{G}_{r,S} = \mathbf{G}_{r,S} \setminus \mathbf{G}_{r,S}^\circ$.

Definition 4.6 (Matroid)**.** A matroid $M = (S, \mathcal{B})$ consists of a finite set S and a non-empty collection $\mathcal{B}$ of subsets of S, called bases, such that for any two bases B, B' and for all $b \in B \setminus B'$ there exists a $b' \in B' \setminus B$ such that $(B \cup \{b'\}) \setminus \{b\}$ is a basis.

Remark 4.7. The cardinality of all bases is the same, it is the rank of a matroid ([**Oxl92**, Lemma 1.2.1]).

There is a stratification of $\mathbf{G}_{r,S}$ by matroid cells (see [**BLS**+**99**, Chapter 2.4] or [**GGMS88**]):

Definition 4.8 (Matroid Cells of the Grassmannian)**.**

(1) Let $p \in \mathbf{G}_{r,S}$. Define a matroid M_p on S via the following rule:
B is a base of M_p if $p_B \ne 0$.
(2) If M is any rank r matroid on S define the realization space of M as
$$\mathcal{R}_\mathbb{K}(M) := \{p \in \mathbf{G}_{r,S}(\mathbb{K}) \mid M_p = M\}.$$

Hence $\mathbf{G}_{r,S}$ is equal to the disjoint union $\bigcup_M \mathcal{R}_\mathbb{K}(M)$ where M ranges over all matroids of rank r.

Let us consider the tropicalization of the classical Grassmannian.

Definition 4.9 (Tropical Grassmannian)**.** Let $\mathbb{K}$ be an algebraically closed field with a surjective valuation $\mathrm{val} : \mathbb{K}^\times \to \mathbb{R}$ and let $\mathbb{k}$ be the residue field.

The tropical Grassmannian is the set
$$\mathbf{G}_{r,S}^{(\mathbb{K},\mathbb{k})}(\mathbb{T}) := \operatorname{trop} \mathbf{G}_{r,S}(\mathbb{K}) \subseteq \mathbb{P}\left(\mathbb{T}^{\binom{S}{r}}\right).$$

Our main focus will be on the case where $\mathbb{K}$ is of characteristic zero with residue field $\mathbb{C}$, hence we abbreviate
$$\mathbf{G}_{r,S}(\mathbb{T}) := \mathbf{G}_{r,S}^{(\mathbb{C}((t^\mathbb{R})),\mathbb{C})}(\mathbb{T}) := \operatorname{trop} \mathbf{G}_{r,S}(\mathbb{C}((t^\mathbb{R}))) \subseteq \mathbb{P}\left(\mathbb{T}^{\binom{S}{r}}\right).$$

It is the closure of a balanced polyhedral fan of dimension $(|S|-r)r$ in the torus of $\mathbb{P}\left(\mathbb{T}^{\binom{S}{r}}\right)$:
$$\mathbf{G}_{r,S}^{\circ}(\mathbb{T}) := \left\{ x \in \mathbf{G}_{r,S}(\mathbb{T}) \mid x_I \neq -\infty \text{ for all } I \in \binom{S}{r} \right\} \subseteq \mathbb{R}^{\binom{S}{r}}/\mathbb{R}$$

Remark 4.10. By Theorem 3.4 the tropicalization depends only on the characteristic pair $(\operatorname{char} \mathbb{K}, \operatorname{char} \mathbb{k})$ and not on the actual fields. The tropical Grassmannian was introduced in the paper "The Tropical Grassmannian" [SS04]. Two important results are:

- For $r = 2$ the Plücker relations form a tropical basis (independent of any characteristic) and we have the description
$$\mathbf{G}_{2,n}^{(\mathbb{K},\mathbb{k})}(\mathbb{T}) = \mathbf{G}_{2,n}(\mathbb{T}) = \bigcap_{i<j<k<l} \mathcal{T}(x_{ij} \odot x_{ik} \oplus x_{ik} \odot x_{jl} \oplus x_{il} \odot x_{jk}).$$

- There is a dependence upon the characteristic pair $(\operatorname{char}\mathbb{K}, \operatorname{char}\mathbb{k})$ for $r \geq 3, |S| \geq 7$.

In [SS04] several related notions of tropical Grassmannian are considered. The authors do not discuss an equivalent of $\mathbf{G}_{r,n}(\mathbb{T})$, the set $\mathbf{G}_{r,n}^{\circ}(\mathbb{T})$ appears as $\mathcal{G}'_{r,n}$.

Matroids impose the same decomposition on the tropical Grassmannian as on the classical Grassmannian. We have
$$\begin{aligned}
\mathcal{R}_\mathbb{T}(M) &:= \operatorname{trop} \mathcal{R}_\mathbb{K}(M) \\
&= \{\operatorname{trop} p \in \mathbf{G}_{r,n}(\mathbb{T}) \mid M_p = M\} \\
&= \{p \in \mathbf{G}_{r,n}(\mathbb{T}) \mid p_B = -\infty \text{ iff } B \text{ is not a base of } M\}
\end{aligned}$$
for a rank r matroid on S.

We will call a subset $\mathcal{R}_\mathbb{T}(M) \subseteq \mathbf{G}_{r,S}(\mathbb{T})$ a matroid stratum of the tropical Grassmannian in analogy to the classical situation.

Definition 4.11 (Tropical Plücker Vector). A point $p \in \mathbb{P}\left(\mathbb{T}^{\binom{S}{r}}\right)$ that satisfies the three-term Plücker relations

- the maximum in $\max(p_{Eij} + p_{Ekl}, p_{Eik} + p_{Ejl}, p_{Eil} + p_{Ejk})$ is achieved at least twice for all $E \in \binom{S}{r-2}$, i, j, k, l distinct and in $S \setminus E$

is called a tropical Plücker vector of rank r.

Every tropical Plücker vector p defines a rank r matroid M_p on S via the rule: I is a base of M_p if and only if $p_I \neq -\infty$.

Remark 4.12. The set of all rank r tropical Plücker vectors with finite coordinates $\mathcal{T}_{r,n}$ is called the set of r-trees in [**SP05**, Chapter 3]. All points in the tropical Grassmannian are tropical Plücker vectors, in general it is a proper subset. However, the tropical Grassmannian $\mathbf{G}_{2,S}(\mathbb{T})$ is the set of all rank 2 tropical Plücker vectors as it is cut out by the three-term Plücker relations.

Definition 4.13. Let p be a tropical Plücker vector of rank r on $[n]$. We define the set $L(p) \subseteq \mathbb{P}(\mathbb{T}^n)$ as follows

$$L(p) := \bigcap_{1 \leq j_1 < \cdots < j_{r+1} \leq n} \left\{ \max_{i=1,\ldots,r+1} \{p_{j_1 \cdots \hat{j_i} \cdots j_{r+1}} + x_{j_i}\} \text{ is achieved twice} \right\} \subseteq \mathbb{TP}^{n-1}$$

Any subset of $\mathbb{TP}^{n-1}$ of this form is called a tropical linear space of dimension $r-1$. A tropical line is a tropical linear space of dimension one.

Remark 4.14. $L(p)$ is a balanced polyhedral complex with all weights equal to one, but we do not use this fact for general tropical linear spaces. In the case of lines it follows since tropical lines are tropicalizations from ordinary linear spaces.

Remark 4.15. This seems to be the most general definition of a tropical linear space so far. It includes the tropical linear spaces of [**Spe08**], where Speyer only considered tropical Plücker vectors with finite coordinates.

This definition also includes Bergman fans of matroids [**Stu02**, Section 9.3], which correspond to tropical Plücker vectors whose entries are all either zero or $-\infty$ (the so-called constant-coefficient case).

The idea of [**Spe08**] is to create linear spaces that are polyhedral complexes and not just fans by considering a regular subdivision of the k-th hypersimplex induced by a tropical Plücker vector. All these linear spaces have the same recession fan, the k-skeleton of the fan of projective space. The k-th hypersimplex is the matroid polytope of the uniform rank k matroid.

The Bergman fan of a matroid M is a subfan of the normal fan of matroid polytope of the dual matroid M^*.

Definition 4.13 combines these notions:
A tropical Plücker vector defines both a matroid (via the infinite coordinates) and a regular subdivision of the matroid polytope (via the finite coordinates).

Theorem 4.16. *Let $\mathbb{K}$ be an algebraically closed field with a surjective non-Archimedean valuation.*

- *Tropical lines are in one-to-one correspondence with points in the tropical Grassmannian* $\mathbf{G}_{2,n}(\mathbb{T})$.
- *If L is a linear subspace of $\mathbb{P}^{n-1}$ corresponding to a point $p \in \mathbf{G}_{2,n}(\mathbb{K})$ then* $\operatorname{trop} L$ *is tropical line corresponding to the point* $\operatorname{trop} p$ *in the tropical Grassmannian.*
- *A tropical line in $\mathbb{TP}^{n-1}$ is the tropicalization of a line in $\mathbb{KP}^{n-1}$.*

PROOF.

- Points in the tropical Grassmannian $\mathbf{G}_{2,n}(\mathbb{T})$ are the same as tropical Plücker vectors of rank two. Therefore there is a surjective map from points in the Grassmannian to tropical lines. Points in different matroid strata lead to lines with different rays, so all we need to show is that different points within one matroid stratum produce different lines. By the result of Lemma 4.27 points in a matroid stratum uniquely encode an abstract metric tree and translation in some subtorus of $\mathbb{TP}^{n-1}$ and hence encode different tropical lines.
- The tropical polynomials cutting out $L(\operatorname{trop} p)$ are exactly the tropicalizations of the polynomials cutting out $L(p)$, hence $\operatorname{trop} L(p) = L(\operatorname{trop} p)$.
- Let $L = L(p)$ be a tropical line. The tropical Plücker vector p lies in $\mathbf{G}_{2,n}(\mathbb{T})$, so there is a point P in $\mathbf{G}_{2,n}(\mathbb{K})$ with $p = \operatorname{trop} P$. But then $L(p) = L(\operatorname{trop} P) = \operatorname{trop} L(P)$ is the tropicalization of a line over $\mathbb{K}$.

□

Let L be a line in $\mathbb{P}(\mathbb{K}^S)$ and g a point in the torus T of $\mathbb{P}(\mathbb{K}^S)$. The element g acts on points of $\mathbb{P}(\mathbb{K}^S)$ by coordinate-wise multiplication. Then $g \cdot L$ is again a linear space: If $L = \operatorname{span}(b_1, b_2)$ then $g \cdot L = \operatorname{span}(g \cdot b_1, g \cdot b_2)$. Thus $L = L(b_1 \wedge b_2)$ and $g \cdot L = (g \cdot b_1 \wedge g \cdot b_2)$.

We can define an action of T on $\mathbf{G}_{2,S}(\mathbb{K})$ via

$$g \cdot p := (g_i \cdot g_j \cdot p_{ij})_{ij}$$

This action satisfies $L(g \cdot p) = g \cdot L(p)$.

We can do the same thing tropically: Let $L = L(p)$ be a line in $\mathbb{P}(\mathbb{T}^S)$ and g in the torus $\mathbb{R}^S/\mathbb{R}$ of $\mathbb{P}(\mathbb{T}^S)$. We define

$$g + p := (g_i + g_j + p_{ij})_{ij}$$

Now let $x \in L(p)$. That means the maximum in $p_{ij} + x_k \oplus p_{ik} + x_j \oplus p_{jk} + x_i$ is achieved twice. But then the maximum in

$$p_{ij} + g_i + g_j + g_k + x_k \oplus p_{ik} + g_i + g_k + g_j + x_j \oplus p_{jk} + g_j + g_k + g_i + x_i$$

is also achieved twice, hence $g + x \in L(g \cdot p)$.

This operation obviously commutes with tropicalization:

trop $L(g \cdot p) = L(\text{trop}(g \cdot p)) = L(\text{trop } g + \text{trop } p) = \text{trop } g + L(\text{trop } p)$.

We let $\mathrm{M}_{0,S}(\mathbb{T})$ be the quotient of the tropical Grassmannian by this operation:

Definition 4.17. We set
$$\mathrm{M}_{0,S}(\mathbb{T}) := \mathbf{G}^{\circ}_{2,S}(\mathbb{T}) / \left(\mathbb{R}^S / \mathbb{R}\right)$$
and abbreviate
$$\mathrm{M}_{0,n}(\mathbb{T}) := \mathrm{M}_{0,[n]}(\mathbb{T}) = \mathbf{G}^{\circ}_{2,n}(\mathbb{T}) / \left(\mathbb{R}^n / \mathbb{R}\right).$$

Remark 4.18. If we look at the tropical Grassmannian $\mathbf{G}^{\circ}_{2,n}(\mathbb{T})$, then it is a fan with an $(n-1)$-dimensional lineality space. This lineality space is the orbit of 0 in $\mathbf{G}^{\circ}_{2,n}(\mathbb{T})$ under the action of the torus of $\mathbb{TP}^{n-1}$. We can describe it as
$$\{[(a_i + a_j)_{ij}] \mid a = (a_1, \ldots, a_n) \in \mathbb{R}^n\}.$$
So $\mathrm{M}_{0,n}(\mathbb{T})$ is just the quotient of the fan $\mathbf{G}^{\circ}_{2,n}(\mathbb{T})$ by its lineality space.

In [SS04] this space $\mathrm{M}_{0,n}(\mathbb{T})$ appears as $\mathcal{G}''_{2,n}$. It is also discussed in [GM07, Section 5] and [GKM09, GM08] (where the authors call it $\mathcal{M}_n$).

We can generalize a certain aspect of this construction to arbitrary varieties:

Proposition 4.19. *Let $T = (\mathbb{K}^{\times})^n$ be the n-dimensional algebraic torus where $\mathbb{K}$ is an algebraically closed field with a non-Archimedean valuation. Let C be an irreducible subvariety of T and H a subtorus of T that acts on C. Then the lineality space of $\text{trop } C$ contains $\text{trop } H$.*

PROOF. Let c be a point of C. Then $\text{trop } Hc$ is equal to the set of translates $\text{trop } c + \text{trop } H$ where $\text{trop } H$ is a real vector space. Hence $\text{trop } C$ contains $\text{trop } C + \text{trop } H$ and therefore contains H as its lineality space. □

Definition 4.20 (Matroid nomenclature). Let M be a matroid on E.

- An independent set is a subset of E that is contained in a basis.
- A dependent set is subset of E that is not independent.
- A circuit is an inclusion-wise minimal dependent set.
- A loop is an element that is contained in no basis. It is a circuit consisting of one element.
- Two elements a, b of E are parallel if $\{a, b\}$ is a circuit.
- A hyperplane is a proper subset H of E that does not contain a basis, but $H \cup \{s\}$ contains a basis for every s in the complement $E \setminus S$.
- A subset A of E is called an atom, if it contains at least one element b that is contained in a basis and satisfies $a \in A$ if and only if a and b are parallel.
- We let $\text{At}(M)$ be the set of atoms of M.

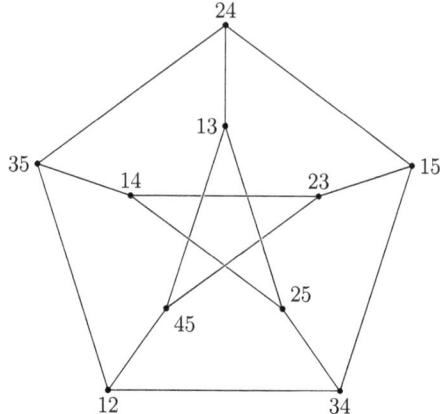

FIGURE 7. Intersecting the fan $M_{0,5}(\mathbb{T})$ with a sphere yields a one-dimensional polyhedral complex that is homeomorphic to the Petersen graph. The 10 vertices correspond to the ten rays indexed by $\binom{[5]}{2}$. They are adjacent if and only if their index sets are disjoint.

In the rank 2 case, atoms and hyperplanes coincide. A rank r matroid has at least r atoms.

Definition 4.21. Let M be a rank r matroid on n elements. We define a tropical Plücker vector e_M via

$$(e_M)_I = \begin{cases} -\infty, & \text{if } I \text{ is a basis of } M \\ 0, & \text{otherwise} \end{cases}$$

We write $L(M)$ for the tropical linear space $L(e_M)$.

Remark 4.22. If M contains no loops then $L(M) \cap \mathbb{R}^n / \mathbb{R}$ is equal to the Bergman fan $\tilde{\mathcal{B}}(M)$ see [**Stu02**, Section 9.3].

Theorem 4.23. [**AK06**, Theorem 3.1] *Cones in the Bergman fan of a matroid M are in one-to-one correspondence with flags of flats of M.*

Theorem 4.24. *Let p be a rank r tropical Plücker vector on n elements. Let e be the tropical Plücker vector defined via:*

$$e_I = \begin{cases} -\infty, & \text{if } p_I = -\infty \\ 0, & \text{otherwise} \end{cases}$$

Then $L(e)$ is the recession fan of $L(p)$.

PROOF. Since e and p have the same matroid the polyhedral complexes $L(e)$ and $L(p)$ are inside the same torus orbit of $\mathbb{TP}(n)$. It is obvious that the polyhedral complex $L(e)$ is a fan. Let r be any ray of a cell of $L(p)$.

That is we have a vertex v and we know that for every point $v + \lambda r$ the maximum in
$$p_{ij} + v_k + \lambda r_k \oplus p_{ik} v_j + \lambda r_j \oplus p_{jk} + v_i + \lambda r_i$$
is achieved twice. That implies that the maximum in
$$0 + \lambda r_k \oplus 0 + \lambda r_j \oplus 0 + \lambda r_i$$
is also achieved twice, hence r is a ray of $L(e)$ and we therefore see that the recession cone of $L(p)$ is a subfan of $L(e)$. Since $L(e)$ is balanced and of degree one we get equality. □

Proposition 4.25. *Let p be a rank r tropical Plücker vector on n elements.*

(1) *If the matroid M_p contains a loop h then $L(p)$ is a subset of the torus divisor $\{x_h = 0\}$ of $\mathbb{TP}(n)$. If we define the tropical Plücker vector p' on n elements via forgetting all cordinates that contain h then $L(p) = L(p')$ when we identify $\mathbb{TP}(n-1)$ with $\{x_h = -\infty\} \subseteq \mathbb{TP}(n)$.*

(2) *Assume that M_p has no loops. Then the rays of $L(p)$ are in bijection with the flats of F as follows: a flat F leads to a ray $r_F = \sum a \in F - e_a$.*

PROOF.

(1) Assume M_p has a loop h. That means h occurs in no basis and $p_{ih} = -\infty$ for all $i \neq h$. Let k and l be two elements from different atoms of M_p. That means $p_{kl} \neq -\infty$. Hence one of the conditions cutting out $L(p)$ is $p_{kh} x_l \oplus p_{lh} x_k \oplus p_{kl} x_h$ which translates to $-\infty \oplus p_{kl} x_h$ and hence to $x_h = -\infty$.

After choosing $x_h = -\infty$ the equations describing $L(p)$ and $L(p')$ are identical.

(2) We know that the cones of $L(p)$ are in bijection with the cones of $L(M_p)$ and hence with flats of flags of M_p. A flat is flag of length one.

Let A be a flat of M_p. Let r be any point in the ray generated by $r_A = \sum a \in A - e_a$. We have to look at the maximum in $p_{ij} + r_k \oplus p_{ik} + r_j \oplus p_{kj} + r_i$ where we can assume $p = e_{M_p}$.

- If $A \cap \{i,j,k\} = \emptyset$ then the maximum is achieved twice since p is a tropical Plücker vector.
- If $A \cap \{i,j,k\} = \{i\}$ then the maximum is achieved twice at $p_{ij} = p_{ik} \geq p_{kj} + r_i$. If we had $p_{ij} \neq p_{ik}$ then that means that either ij or ik are parallel in contradiction to $A \cap \{i,j,k\} = \{i\}$.
- If $A \cap \{i,j,k\} = \{i,j\}$ then the maximum is achieved by $p_{ik} + r_j = p_{kj} + r_i \geq -\infty = p_{ij} + r_k$.
- $A \supseteq \{i,j,k\}$ then all three terms are equal to minus infinity.

Note: If $A \subseteq B$ are two flats then the terms where the minimum is achieved in r_A is a subset of the points where the minimum is achieved in r_B, hence the minimum is achieved twice for $r_A + r_B$, too. This shows that flags of flats lead to cones of $L(p)$.

We have seen that we can factor the generic matroid stratum of the tropical Grassmannian as

$$\mathbf{G}_{2,n}^{\circ}(\mathbb{T}) \approx \mathrm{M}_{0,n}(\mathbb{T}) \times \mathbb{R}^{n-1}$$

where the first factor is the set of "abstract" tropical line in the sense of [**GKM09**], and the second factor determines an embedding to $\mathbb{R}^{n-1} \subseteq \mathbb{TP}^{n-1}$.

We can do this for all matroid strata of $\mathbf{G}_{2,n}(\mathbb{T})$.

Lemma 4.26. *Let M be a matroid on $[n]$ of rank 2 without loops.*

- *There is a piecewise linear homeomorphism $\mathcal{R}_{\mathbb{T}}(M) \cong \mathrm{M}_{0,\mathrm{At}(M)}(\mathbb{T}) \times \mathbb{R}^{n-1}$ if M has at least three atoms.*
- *There is a piecewise linear homeomorphism $\mathcal{R}_{\mathbb{T}}(M) \cong \mathbb{R}^{n-2}$ if M has precisely two atoms.*

PROOF. Let $H_1, \ldots, H_s$ be the atoms of M. After relabeling we can assume that $i \in H_i$.
Let us look at the Plücker relation with indices $i < j < k \leq s < l_0$ Assume that $l_0 \in H_l$

$$p_{ij} + p_{kl_0} \oplus p_{ik} + p_{jl_0} \oplus p_{il_0} + p_{jk}$$

As $\{l, l_0\}$ are parallel, the following Plücker relations

$$p_{ij} + p_{ll_0} \oplus p_{il} + p_{jl_0} \oplus p_{il_0} + p_{jl}$$
$$p_{il} + p_{kl_0} \oplus p_{ik} + p_{ll_0} \oplus p_{il_0} + p_{lk}$$
$$p_{jl} + p_{kl_0} \oplus p_{kl} + p_{jl_0} \oplus p_{ll_0} + p_{jk}$$

degenerate into linear equations

$$p_{il} + p_{jl_0} = p_{il_0} + p_{jl}$$
$$p_{il} + p_{kl_0} = p_{il_0} + p_{lk}$$
$$p_{jl} + p_{kl_0} = p_{kl} + p_{jl_0}$$

If $\{i, l\}$ is not a loop, then we can express this as

$$p_{jl_0} = p_{il_0} + p_{jl} - p_{il}$$
$$p_{kl_0} = p_{il_0} + p_{lk} - p_{il}$$

and transform our original equation into

$$p_{ij} + p_{il_0} + p_{lk} - p_{il} \oplus p_{ik} + p_{il_0} + p_{jl} - p_{il} \oplus p_{il_0} + p_{jk}$$

we can simplify this to

$$p_{ij} + p_{lk} \oplus p_{ik} + p_{jl} \oplus p_{jk} + p_{il}$$

Hence the original relation was superfluous and we only need to consider the Plücker relations with indices from the set $[s]$. $\square$

Lemma 4.27. *Let M be a matroid on $[n]$ of rank 2 with m loops.*

- *If M has $k \geq 3$ atoms then $\mathcal{R}_{\mathbb{T}}(M) \cong \mathrm{M}_{0,k}(\mathbb{T}) \times \mathbb{R}^{n-m-1}$*
- *If M has no parallel elements then there is a linear isomorphism $\overline{\mathcal{R}_{\mathbb{T}}(M)} \cong \mathbf{G}_{2,n-m}(\mathbb{T})$.*
- *If M has precisely two atoms then $\mathcal{R}_{\mathbb{T}}(M) \cong \mathbb{R}^{n-m-2}$*

PROOF. Assume that $n - m, n - m + 1, \ldots, n$ are the loops of M.

All Plücker relations involving loops degenerate to $-\infty \oplus -\infty \oplus -\infty$. All coordinates p_{ij} involving loops are equal to $-\infty$. That means we have an isomorphism between $\mathcal{R}_{\mathbb{T}}(M)$ and $\mathcal{R}_{\mathbb{T}}(M \setminus \{n - m, \ldots, n\})$ via forgetting all coordinates involving loops. $\square$

CHAPTER 5

Chow Quotients

In this chapter we introduce the notion of Chow quotients and present several results about them. Chow quotients will be an indispensable tool in Chapter 6.

Let F be a balanced fan in a vector space V with lineality space L. For every subspace $W \subseteq L$ we get a balanced fan F/W in V/W.

Now assume that we have a tropical toric variety X with torus V that compactifies F. We would like to find a compactification of F/W. Let $\overline{F}$ be the closure of F in X. The topological quotient $\overline{F}/W \subseteq X/W$ is a compact space, but it is usually not Hausdorff (and the topological quotient X/W is not a toric variety).

The same problem arises in complex algebraic geometry, here F is a subvariety of the torus T of a complex toric variety X. A subtorus H of T is acting on F and on the closure $\overline{F}$. One wants to find a compactification of F/H. The topological quotient F/H is a subvariety of the torus T/H. Again, the quotient $\overline{F}/H$ is not separated in general and X/H is not a toric variety.

In complex algebraic geometry, there are three prominent ways to find a toric variety Y compactifying the torus T/H and the subvariety F/H.

(1) The Mumford quotient or GIT quotient $X/\!\!/_\alpha H$ can be constructed as follows [**MFK94**]: We first choose an equivariant projective embedding $X \subseteq \mathbb{P}(s)$. Choose a map $\alpha : H \to \mathrm{GL}_s$ extending the action of H to $\mathbb{P}(s)$. Now H is acting on the coordinate ring $\mathbb{C}[\mathbb{P}(s)]$ of $\mathbb{P}(s)$ and we can form the invariant subring $\mathbb{C}[\mathbb{P}(s)]^H = \{f \in \mathbb{C}[\mathbb{P}(s)] \mid hf = f \text{ for all } h \in H\}$. The quotient $X/\!\!/_\alpha H$ is then defined as $\mathrm{Proj}\,\mathbb{C}[\mathbb{P}(s)]^H$.

   The GIT quotient of a projective toric variety is again a projective toric variety [**KSZ91**, Prop. 3.2].

(2) The Chow quotient $X/\!\!/ H$ is the closure of T/H in the Chow variety of X. Points in the Chow quotient correspond to effective cycles formed by H-orbits of points in X.

   The Chow quotient of a projective toric variety by a subtorus is again a projective toric variety, it is described by the secondary polytope of the polytope of X. The corresponding fan is the secondary fan [**KSZ91**, Prop. 2.3].

(3) The Hilbert quotient $X/\!\!/_{\mathrm{Hi}} H$ is the closure of T/H in the Hilbert scheme of X [**Kol96**, Claim 1.8.2]. Points in the Hilbert quotient correspond to subschemes formed by H-orbits of points in X.

The Hilbert quotient of a projective toric variety by a subtorus is again a projective toric variety, it is described by the state polytope of the ideal of X. The corresponding fan is the Gröbner fan of $I(X)$.

These quotients are related as follows (a discussion of the geometric aspects of these quotients can be found in [**Hu05**]):

Theorem 5.1.

(1) *There is a regular birational map from the Hilbert quotient $X /\!/_{\text{Hi}} H$ to the Chow quotient $X /\!/ H$* [**Kap93**, Statement (0.5.9)].
(2) *There is a regular birational map from the Chow quotient $X /\!/ H$ to any GIT quotient $X /\!/_\alpha H$* [**KSZ91**, Prop. 3.3].
(3) *The Chow quotient $X /\!/_{\text{Hi}} H$, the Hilbert quotient $X /\!/_{\text{Hi}} H$ and every GIT quotient $X /\!/_\alpha H$ can be realized as a GIT quotient $Y /\!/_\beta G$ for suitable choices of Y, G and β* [**GM07**, Prop 4.3, Theorem 4.6].

We will begin with studying the Chow variety and follow the treatment detailed in [**GKZ94**]. A more modern and scheme-theoretic approach can be found in [**Kol96**].

Definition 5.2 (Chow Form). Let $X \subseteq \mathbb{CP}^{n-1}$ be an irreducible subvariety of dimension $k-1$ and degree d.

Let $\mathcal{Z}(X)$ be the set of all $(n-k-1)$-dimensional linear subspaces of $\mathbb{CP}^{n-1}$ that intersect X. Then $\mathcal{Z}(X)$ is an irreducible hypersurface of degree d in $\mathbf{G}_{n-k,n}(\mathbb{C})$ defined by a single homogeneous polynomial R_X [**GKZ94**, Prop 2.1,2.2]. That polynomial is the Chow form of X (unique up to a scalar multiple).

Example 5.3. If $X = \{x \in \mathbb{P}^{n-1} | f(x) = 0\}$ is a hypersurface, then $n - (n-1) - 1 = 0$ and $\mathcal{Z}(X)$ is the set of all points intersecting X. That means $X = \mathcal{Z}(X)$ and $R_X = f$ under the identification $\mathbf{G}_{n-1,n} = \mathbb{P}^{n-1}$.

An irreducible subvariety Z of dimension k is uniquely determined [**GKZ94**, Prop. 2.5] by the coefficients of its Chow form in $\mathbb{K}[\mathbf{G}_{n-k,n}]$. We define the Chow form of an effective cycle of dimension k to be the product of the Chow forms of its irreducible components (with multiplicities). Thus an effective cycle of dimension k and degree d is uniquely determined by its Chow form in $\mathbb{K}[\mathbf{G}_{n-k,n}]_d$. This allows us to turn the set of all these effective cycles into an algebraic variety.

Definition 5.4 (Chow Variety). Let X be a projective variety with a fixed projective embedding. The set of all effective k-cycles on X can be given the structure of a projective variety $\mathcal{C}_k(X)$, called the Chow variety of X of dimension k. Its irreducible components are given by the subsets of cycles of a fixed homology class δ, we denote that variety by $\mathcal{C}_k(X, \delta)$.

Example 5.5. We already know the Chow variety of k-cycles in $\mathbb{P}^n$ that have the same homology class as a k-plane h. It is the set of all k-dimension linear subspaces of $\mathbb{P}^n$ and

$$\mathcal{C}_k(\mathbb{P}^n, [h]) = \mathbf{G}_{k+1, n+1}.$$

Let X be a toric variety with torus T and H a subtorus. We will now use the Chow variety to construct a compactification $X /\!\!/ H$ of the quotient T/H.

Definition 5.6 (Chow Quotient). Let X be a projective variety and G a reductive algebraic group acting on X. Let U be a dense open subset of X such that the orbit closure $\overline{Gx}$ has the same dimension k and homology class δ for every point x of U. This defines an inclusion $U/G \to \mathcal{C}_k(X, \delta)$.

The Chow quotient of X by G is defined as the closure of U/G in $\mathcal{C}_k(X, \delta)$. It is independent of the choice of U [**Kap93**, Remark 0.1.8] and denoted $X /\!\!/ G$.

Remark 5.7. The Hilbert quotient is defined similarly as the closure of U/G in the Hilbert scheme.

If $X = \mathbf{X}_F$ is a toric variety and G is a subtorus of the torus T of X, then the Chow quotient $X /\!\!/ G$ is again a toric variety. We will now describe the corresponding fan.

Definition 5.8 (Quotient Fan). Let F be a rational fan in a vector space V and W a subspace. Let σ° denote the relative interior of a cone σ. We define an equivalence relation on V/W as follows: Two points $x + W$ and $\tilde{x} + W$ are equivalent if they meet the same cones, i.e. $x + W \sim \tilde{x} + W$ if and only if

$$\{\sigma \in F \mid x + W \cap \sigma^\circ \neq \emptyset\} = \{\tilde{\sigma} \in F \mid \tilde{x} + W \cap \tilde{\sigma}^\circ \neq \emptyset\}.$$

The equivalence classes of this relation are relative interiors of cones and define a fan. We call it the quotient fan F/W.

Remark 5.9. Each equivalence class $[x + W]$ is equal to the multi-intersection

$$[x + W] = \bigcap_{x + W \cap \sigma^\circ \neq \emptyset} \sigma + W$$

Theorem 5.10. [**KSZ91**, Prop. 2.3] *Let X be a normal projective toric variety and H a subtorus of the torus of X. Then $X /\!\!/ H$ is a projective toric variety. Its fan is the quotient fan of the fan of X.*

If $F = \mathcal{N}(P)$ is the normal fan of a polytope, then F/W is also the normal fan of a polytope. We will now construct this polytope.

Definition 5.11. [**GKZ94**, Chapter 7.2] A marked polytope is a pair (Q, A) where Q is a polytope and A is a finite family of points such that $Q = \text{conv } A$. A subdivision of (Q, A) is a family $S = \{(Q_i, A_i) \mid i \in I\}$ of marked polytopes such that

(1) each A_i is a subset of A and $\dim(Q_i) = \dim(Q)$,
(2) any intersection $Q_i \cap Q_j$ is a face (possibly empty) of both Q_i and Q_j and
$$A_i \cap (Q_i \cap Q_j) = A_j \cap (Q_i \cap Q_j),$$
(3) the union of all Q_i coincides with Q.

Example 5.12. Let $A \subseteq \mathbb{R}^n$ be a finite family and $Q = \operatorname{conv} A$. Let $\psi : A \to \mathbb{R}$ be any function and define the polyhedron
$$G_\psi = \operatorname{conv}\{(\omega, y) \in \mathbb{R}^n \times \mathbb{R} \mid y \leq \psi(\omega)\}.$$
The piecewise-linear function $g_\psi : Q \to \mathbb{R}$ is defined as
$$g_\psi(x) = \max\{y \in \mathbb{R} \mid (x, y) \in G_\psi\}.$$
Let the Q_i be the projections of the bounded faces of G_ψ of codimension one. Let A_i consist of all $\omega \in A \cap Q_i$ such that $g_\psi(\omega) = \psi(\omega)$. Then $\{(Q_i, A_i)\}$ forms a subdivision of (Q, A) that is denoted $S(\psi)$.

If the same point p occurs multiple times in the family A, then the function ψ may take different values each time. Only those instances with maximal ψ-value occur in the subdivision $S(\psi)$.

Definition 5.13. A subdivision S of (Q, A) is called regular if it is of the form $S(\psi)$ for some $\psi \in \mathbb{R}^A$.

Definition 5.14 (Fiber Polytope). Let $\pi : P \to Q$ be a map of polytopes, i.e. P is a polytope, π a linear map and $Q = \pi(P)$. A section of π is a continuous map $\gamma : Q \to P$ such that $\pi \circ \gamma = \operatorname{id}_Q$. The fiber polytope of P and Q is defined as the set of all component-wise integrals of such sections:
$$\Sigma(P, Q) := \left\{ \frac{1}{\operatorname{Vol}(Q)} \int_Q \gamma(t) \mathrm{d}t \;\middle|\; \gamma \text{ is a section of } \pi \right\}.$$

If Q is a polytope with n vertices, then there is a canonical map from the simplex with n vertices Δ_n to Q. This special case of a fiber polytope is called a secondary polytope and denoted $\Sigma(Q) := \Sigma(\Delta_n, Q)$.

Theorem 5.15. Let $P \subseteq V$ be a polytope, $\pi : V \to W$ a linear map and $Q = \pi(P)$.

(1) The fiber polytope is a polytope. Its dimension equals $\dim P - \dim Q$.
(2) The points of the secondary polytope of Q represent regular subdivisions of Q, the vertices are in bijection with the regular triangulations of Q.
(3) The normal fan of the fiber polytope equals the quotient fan of normal fan of P:
$\mathcal{N}(\Sigma(P, Q)) = \mathcal{N}(P)/\operatorname{im} \pi^\vee$ where $\pi^\vee : W^\vee \to V^\vee$ is the dual map.

PROOF. [**Zie95**, Theorem 9.6] and [**BS94**, Prop. 2.2] □

Corollary 5.16. *Let H be a subtorus of the torus of $\mathbb{P}^n$. There is a polytope Q such that the toric variety associated to Q equals the closure of H in $\mathbb{P}^n$. Then the Chow quotient $\mathbb{P}^n /\!/ H$ is the toric variety associated to $\Sigma(Q)$, i.e. orbits of $\mathbb{P}^n /\!/ H$ correspond to regular subdivisions of Q.*

Remark 5.17. Since Chow quotients can be expressed entirely in terms of polytopes and fans, we can form Chow quotients of tropical polyhedral complexes inside tropical toric varieties by vector spaces acting on the variety. If we have an algebraic subvariety inside a toric variety, then the Chow quotient of tropicalization is tropicalization of the Chow quotient.

Example 5.18. Let us show that the secondary polytope of a pentagon is again a pentagon. There are five ways to triangulate a pentagon, hence the secondary polytope has five vertices. The pentagon has five vertices, so the secondary polytope is the fiber polytope of a four-dimensional simplex to the two-dimensional pentagon and therefore two-dimensional. A two-dimensional polytope with five vertices must necessarily be a pentagon.

Lemma 5.19. *Let X and Y be projective toric varieties and assume the torus H is acting on both X and Y. Let H act on $X \times Y$ via $h(x,y) = (hx, hy)$. Then there are maps*

$$(X \times Y) /\!/ H \to X /\!/ H$$

and

$$(X \times Y) /\!/ H \to Y /\!/ H.$$

PROOF. Assume we have $X = \mathbf{X}_F$ and $X /\!/ H = \mathbf{X}_{F/L}$, $Y = \mathbf{X}_G$ and $Y /\!/ H = \mathbf{X}_{G/K}$. Let V be the diagonal embedding of $L \cong K$ into $\text{span}(|F| \times |G|)$. Then $X \times Y = \mathbf{X}_{F \times G}$ and $(X \times Y) /\!/ H = \mathbf{X}_{F \times G/V}$.

We need to show that the projection pr maps every cone σ of $F \times G/V$ into a cone τ of F/L. We have a representation $\sigma = \bigcap (\sigma_i^1 \times \sigma_i^2)/V$.

This gets mapped to $\bigcap \sigma_i^1/L$. We need to show that this is a cone of F/L. Assume not. That means there is a cone $\tilde{\sigma}^1/L$ that intersects $\bigcap \sigma_i^1/L$. But then $\tilde{\sigma}^1 \times |G|/V$ intersects $\sigma = \bigcap(\sigma_i^1 \times \sigma_i^2)/V$, and therefore there is a cone $\tilde{\sigma}^1 \times \tilde{\sigma}^2/V$ that intersects σ. Hence σ was not a cone of $F \times G/L$. □

Theorem 5.20. *Let A be a $d \times n$-matrix. Let T be the subtorus of $\mathbb{P}(n)$ whose character lattice is the integer row space of A. Let $\text{conv}\, A$ be the convex hull of the columns of A. Then*

$$\dim \overline{Te_0} = \dim \text{conv}\, A$$

and

$$\deg \overline{Te_0} = \text{Vol}(\text{conv}\, A).$$

65

PROOF. This was proven in [**Stu95**, Lemma 4.2, Theorem 4.16] as well as in [**MMKZ92**, Prop. 1.1, Cor. 5.4]. □

Lemma 5.21. [**BS94**, Cor. 2.5] *Let X be a projective toric variety with torus T and $G \subseteq H \subseteq T$ subtori. Then there is a map*

$$(X /\!/ G) /\!/ (H/G) \to X /\!/ H.$$

Lemma 5.22. *Let $x_1, \ldots, x_{n+1}$ be coordinates on the character lattice $\mathbb{Z}^{n+1}/\mathbb{Z} \cong \mathbb{Z}^n$ of $\mathbb{P}(n+1)$. Let H_1 be the subtorus of $T(n+1)$ defined via $H_1 = \mathrm{span}(x_{n+1}) \otimes \mathbb{K}^\times$. Then*

$$\mathbb{P}(n+1) /\!/ H_1 \cong \mathbb{P}(n).$$

PROOF. We need to study the fiber polytope $\Sigma(\Delta_{n+1}, I)$ where I is the unit interval, the convex hull of zero and one. Every vertex corresponds to a tight subdivision, i.e. one which contains vertex zero of I exactly once and vertex one exactly once. There are n vertices of Δ_{n+1} which map to zero and exactly one which maps to one, so we have a polytope of dimension $n-1$ with n vertices. Thus $\Sigma(\Delta_{n+1}, I)$ is a simplex and $\mathbb{P}(n+1) /\!/ H_1 \cong \mathbb{P}(n)$. □

Lemma 5.23. *Let $x_1, \ldots, x_{n+1}$ be coordinates on the character lattice $\mathbb{Z}^{n+1}/\mathbb{Z} \cong \mathbb{Z}^n$ of $\mathbb{P}(n+1)$. Let H_1 be the subtorus of $T(n+1)$ defined via $H_1 = \mathrm{span}(x_{n+1}) \otimes \mathbb{K}^\times$ and consider it as a subtorus of $T\binom{n+1}{2}$. Then*

$$\mathbb{P}\binom{n+1}{2} /\!/ H_1 \cong \mathbb{P}\binom{n}{2} \times \mathbb{P}(n).$$

PROOF. We need to study the fiber polytope $\Sigma(\Delta_{\binom{n+1}{2}}, I)$ where I is the unit interval, the convex hull of zero and one. Every vertex corresponds to a tight subdivision, i.e. one which contains vertex zero of I exactly once and vertex one exactly once. There are $\binom{n}{2}$ vertices of $\Delta_{\binom{n+1}{2}}$ which map to zero and n which map to one (it is $n + \binom{n}{2} = \binom{n+1}{2}$). We can choose these independently and get therefore $n \cdot \binom{n}{2}$ vertices of $\Sigma(\Delta_{\binom{n+1}{2}}, I)$. This means $\Sigma(\Delta_{\binom{n+1}{2}}, I)$ is combinatorially isomorphic to $\Delta_n \times \Delta_{\binom{n}{2}}$. Hence $\mathbb{P}\binom{n+1}{2} /\!/ H_1 \cong \mathbb{P}\binom{n}{2} \times \mathbb{P}(n)$. □

Corollary 5.24.

$$\mathbf{G}_{2,n+1} /\!/ H_1 \cong F\mathbf{G}_{2,n}.$$

PROOF. Let us look at $\mathbf{G}_{2,n+1}$. It is cut out by the Plücker relations $p_{ij}p_{kl} - p_{ik}p_{jl} + p_{il}p_{jk}$ with $1 < i < j < k < l$ and the additional conditions $p_{1j}p_{kl} - p_{1k}p_{jl} + p_{1l}p_{jk}$ with $1 < j < k < l$. If we use $p_{12}, p_{13}, \ldots, p_{1,n+1}$ as homogeneous coordinates on $\mathbb{P}(n)$ then these are exactly the equations describing $F\mathbf{G}_{2,n}$. Hence $\mathbf{G}^\circ_{2,n+1}/H_1 = F\mathbf{G}_{2,n} \cap T$ where T is the maximal torus of $\mathbb{P}\binom{n}{2} \times \mathbb{P}(n)$. The statement then follows with the previous lemma. □

We will now focus on a special case of Chow quotients: quotients of $\mathbb{P}\binom{n}{k}$ by a torus H that acts on the Grassmannian, i.e. $H \cdot \mathbf{G}_{k,n} \subseteq \mathbf{G}_{k,n}$.

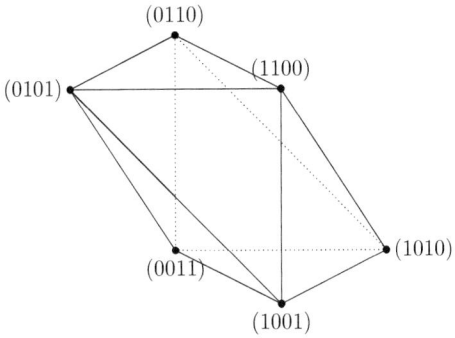

FIGURE 8. The second hypersimplex Δ_4^2 is an octahedron. It is the matroid polytope of the uniform rank two matroid on four elements.

Let $A = (A_I)_{I \in \binom{[n]}{k}}$ be a family of $\binom{n}{k}$ vectors in $\mathbb{Z}^d$. Then A can be used to define a map $(\mathbb{K}^\times)^d \to (\mathbb{K}^\times)^{\binom{n}{k}}/\mathbb{K}^\times \subseteq \mathbb{P}\binom{n}{k}$. Thus A defines a d-dimensional subtorus H_A of $\mathbb{P}\binom{n}{k}$. The orbits of the Chow quotient $\mathbb{P}\binom{n}{k} /\!/ H_A$ are enumerated by all subdivisions of the marked polytope $P = \operatorname{conv} A$. The torus $T(n)$ of $\mathbb{P}(n)$ acts on $\mathbf{G}_{k,n}$ and on $\mathbb{P}\binom{n}{k}$. Let us assume that H_A is a subtorus of of $T(n)$ and acts on $\mathbf{G}_{k,n}$. We want to look at a special kind of subdivisions of P:

Definition 5.25. Let $A = (A_I)_{I \in \binom{[n]}{k}}$ be a family of $\binom{n}{k}$ vectors in $\mathbb{Z}^d$ such that H_A acts as a subtorus of $T(n)$ on $T\binom{n}{k}$. Let M be a rank k matroid on $[n]$. Then M defines a marked polytope $\operatorname{conv}\{A_B \mid B \text{ is a basis of } M\}$. We call marked polytopes of this form matroid polytopes with respect to A.

A matroid decomposition with respect to A is a subdivision of $\operatorname{conv} A$ by matroid polytopes with respect to A.

A matroid polytope is called realizable over $\mathbb{K}$ if the corresponding matroid is realizable over $\mathbb{K}$ and a matroid decomposition is realizable over $\mathbb{K}$ if it is a decomposition with realizable matroid polytopes.

Remark 5.26. This generalizes the notions of matroid polytope from [**GGMS88**] and matroid decomposition from [**Kap93**].

To be compatible with the terminology of [**GGMS88**] and subsequent works, a matroid polytope (without further specification) is a matroid polytope with respect to the k-th hypersimplex where k is the rank of the matroid (see Figure 8 and Figure 9 on the following page for an example).

We have the following reason to study these generalized matroid polytopes:

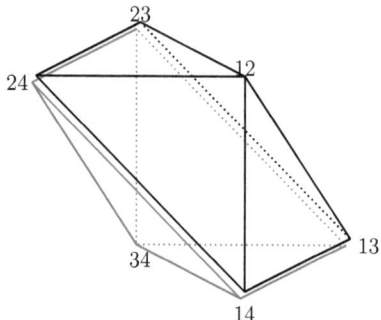

Matroid M_{12} has 1 and 2 as parallel elements.
Matroid M_{34} has 3 and 4 as parallel elements.

FIGURE 9. A matroid decomposition of the second hypersimplex Δ_4^2 into two pyramids

Proposition 5.27. *Let $A = (A_I)_{I \in \binom{[n]}{k}}$ be a family of $\binom{n}{k}$ vectors in $\mathbb{Z}^d$ and assume that H_A acts equivariantly on $\mathbf{G}_{k,n}(\mathbb{K})$. Then $\mathbf{G}_{k,n}(\mathbb{K}) /\!/ H_A$ intersects a torus orbit $O(\sigma)$ of $\mathbb{P}\binom{n}{k} /\!/ H_A$ if and only if the subdivision corresponding to σ is a realizable matroid subdivision.*

PROOF. Let Z be a point in $O(\sigma) \cap \mathbf{G}_{k,n} /\!/ H_A$. Then there is a sum of cycles $Z = \sum c_i \overline{H_A x^{(i)}}$ with $x^{(i)} \in \mathbf{G}_{k,n}$. Combining Corollary 5.16 with [**LJBS90**, Lemma 4.3], we see that the orbit $O(\sigma)$ containing Z corresponds to the subdivision

$$\operatorname{conv} A = \bigcup_i \operatorname{conv}\left\{ A_I \mid x_I^{(i)} \neq 0 \right\}.$$

On the other hand, $x^{(i)}$ lies in a stratum $\mathcal{R}(M_i)$ corresponding to a realizable matroid M_i. Hence each polytope $\operatorname{conv}\{A_I \mid x_I^{(i)} \neq 0\}$ is a realizable matroid polytope.

Now let σ be a realizable matroid decomposition with matroids M_i. Let x be a generic point in $\mathbf{G}_{k,n}$ and let $x^{(i)}$ be the limit of x in $\mathcal{R}(M_i)$. We can form the cycle $Z = \sum c_i \overline{H_A x^{(i)}}$ with $c_i = [(\operatorname{span}(A) + \sigma_i) \cap \mathbb{Z}^d : (\operatorname{span}(A) \cap \mathbb{Z}^d) + (\sigma_i \cap \mathbb{Z}^d)]$. Then Z is a point of $\mathbb{P}\binom{n}{k} /\!/ H_A$ by [**KSZ91**, Prop. 1.1]. It is a point of $\mathbf{G}_{k,n} /\!/ H_A$ by [**Hu05**, Theorem 3.13]. □

Remark 5.28. This result was already used in [**Kap93**]. We will only be interested in the rank two case, where all matroids are realizable.

CHAPTER 6

Rational Tropical Curves

In this chapter we combine our knowledge of the compact tropical Grassmannian and Chow quotients to define a compactification of the spaces of parametrized tropical curves $\mathrm{M}_{0,n}^{\mathrm{lab}}(\mathbb{R}^r, d)$. We then show that this compactification carries a lot of geometric information.

Definition 6.1. Let D be a family of finitely many pairs $(n_i, v_i)_{i \in I}$ where n_i is a positive integer and v_i is a non-zero primitive integral vector in $\mathbb{Z}^r$.

We say that D is a labeled tropical degree if $\sum n_i v_i = 0$.

Let $\mathbb{R}^r$ be the torus of $\mathbb{TP}^r$ and C a one-dimensional balanced polyhedral complex whose rays have directions v_i and weights n_i.

We say that D is a tropical curve of degree D. Note that the family D uniquely determines the projective degree of C, the class of $\overline{C} \subseteq \mathbb{TP}^r$ in $A_1(\mathbb{TP}^r)$.

Let $D(r, d)$ be the degree with members $(1, -e_i)$ for $i = 1, \ldots, r$ and $(1, e_1 + \ldots + e_r)$ each occurring d times where $\{e_i\}_{1 \leq i \leq r}$ is the standard basis of $\mathbb{R}^r$.

If C is a curve in $\mathbb{R}^r$ with degree $D(r, d)$ we say that C has tropical degree d.

Every curve with tropical degree d has projective degree d but not vice versa (see Figure 6 for an example).

The genus of a connected tropical curve C is the first Betti number $\dim H_1(|C|)$ of the underlying topological space, i.e. the number of independent cycles of the graph.

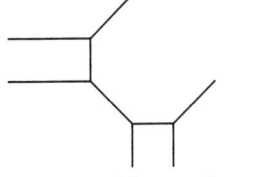

(a) The curve $C_1 = \mathrm{trop}\{x^2 + xy + y^2 + x + y = -1\}$ has tropical degree 2.

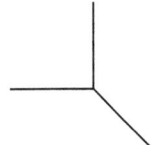

(b) The curve $C_2 = \mathrm{trop}\{xy + y = -1\}$ has projective degree 2 but not tropical degree 2.

FIGURE 10. Two tropical curves with projective degree 2.

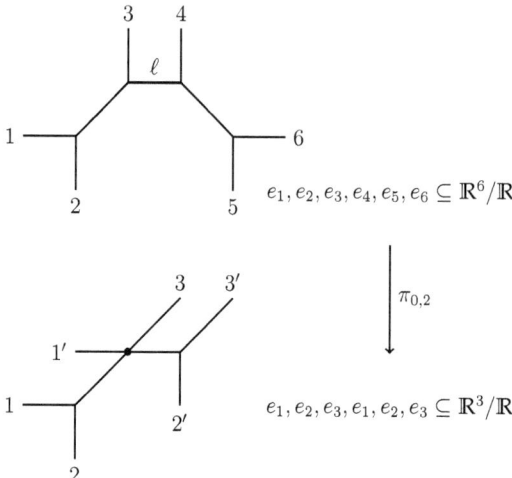

FIGURE 11. A degree 2 curve with a hidden parameter ℓ.

Let D be a tropical degree with pairs (n_i, v_i) with $v_i \in \mathbb{Z}^r$. We define a linear map $\pi_D : \mathbb{R}^{\#D} \to \mathbb{R}^r$ that sends the i-th unit vector e_i to $n_i v_i$. This map is well-defined on $\mathbb{R}^{\#D}/\mathbb{R}$ since $\sum e_i \mapsto \sum n_i v_i = 0$.

Let p be any point in $\mathbf{G}_{2,D}^\circ(\mathbb{T})$. The image $\pi_D(L(p))$ is a connected tropical curve of degree D and parametrized by a tropical curve of genus zero – the image curve $\pi_D(L(p))$ might not be of genus zero.

Let H^D be the kernel of π_D. We know that $\mathbb{R}^D/\mathbb{R}$ acts on $\mathbf{G}_{2,D}^\circ(\mathbb{T})$. If p is a point that maps to the curve C, then so does every point hp for h in H^D. Hence we consider the quotient $\mathbf{G}_{2,D}^\circ/H^D$.

Let $\mathbb{R}^{r-1} \cong \mathbb{R}^r/\mathbb{R}$ be the torus of $\mathbb{TP}^{r-1}$.

Definition 6.2. We define
$$\mathrm{M}_{0,0}^{\mathrm{lab}}(\mathbb{R}^{r-1}, D) := \mathbf{G}_{2,\#D}^\circ(\mathbb{T})/\ker \pi_D$$
to be the parameter space of all tropical curves of degree D that are parametrized by tropical curves of genus zero.

It is possible for such a curve of degree D to have an entire family of preimages, see Figure 11.

There is a straightforward generalization: Let C be a tropical curve of degree D and let $x_1, \ldots, x_n$ be n points that lie on that curve. We define a linear map $\pi_{n,D} : \mathbb{R}^{n+D} \to \mathbb{R}^d$ that sends the j-th unit vector e_j to zero for $j = 1, \ldots, n$ and the i-th unit vector e_i to $n_i v_i$ for

$i = n+1, \ldots, \#D+n$. where $D+n$ stands for the finite family $D \cup [n]$ with $\#D+n$ elements. This map is well-defined on $\mathbb{R}^{D+n}/\mathbb{R}$ since $\sum e_j + \sum e_i \mapsto \sum n_i v_i = 0$.

Let p be any point in $\mathbf{G}^\circ_{2,n+D}(\mathbb{T})$. The image $\pi_{n,D}(L(p))$ is a connected tropical curve of degree D and parametrized by a tropical curve of genus zero. Every ray with direction $-e_j$ of $L(p)$ gets mapped to a point x_j on $\pi_{n,D}(L(p))$.

Definition 6.3. We define
$$\mathrm{M}^{\mathrm{lab}}_{0,n}(\mathbb{R}^{r-1}, D) := \mathbf{G}^\circ_{2,D+n}(\mathbb{T})/\ker \pi_{n,D}$$
to be the parameter space of tropical curves of degree D with n marked points that are parametrized by tropical curves of genus zero.

For every $i = 1, \ldots, n$ there is an evaluation map $\mathrm{ev}_i : \mathrm{M}^{\mathrm{lab}}_{0,n}(\mathbb{R}^{r-1}, D) \to \mathbb{R}^{r-1}$
$$[(p_{kl})] \mapsto [\pi_D((p_{ij})_{j \in D})]$$
that sends a point in $\mathrm{M}^{\mathrm{lab}}_{0,n}(\mathbb{R}^{r-1}, D)$ to the coordinates of the i-th marked point in $\mathbb{R}^r/\mathbb{R}$.

This is basically the setup introduced in [**GM08**]. A more detailed study of this setup can be found in [**GKM09**, Section 4]. Note that the space $\mathrm{M}^{\mathrm{lab}}_{0,n}(\mathbb{R}^{r-1}, D)$ was not defined as $\mathbf{G}^\circ_{2,D+n}(\mathbb{T})/\ker \pi_{n,D}$ but as $(\mathbf{G}^\circ_{2,D+n}(\mathbb{T})/L) \times \mathbb{R}^{r-1}$ where L is the full lineality space of the tropical Grassmannian. This way, however, one loses the embedding of the tropical lines in $\mathbb{R}^r/\mathbb{R}$ and has to recover it using the marked points. This means the setup from [**GM08, GKM09**] required $n > 0$.

Let us look at the complex analogues of these spaces.

Definition 6.4 ($\mathrm{M}_{g,n}$)**.** The moduli space $\mathrm{M}_{g,n} = \mathrm{M}_{g,n}(\mathbb{C})$ is a $3g + n - 3$-dimensional complex variety that parametrizes complex projective nonsingular curves C of genus g together with n distinct marked points $p_1, \cdots, p_n$ on C (we use [**FP97**] as a reference). $\mathrm{M}_{g,n}$ has a compactification $\overline{\mathrm{M}}_{g,n}$ whose points correspond to projective, connected, nodal curves C, together with n distinct, nonsingular, marked points, with a stability condition that is equivalent to the finiteness of automorphism groups.

The distinctive property of $\overline{\mathrm{M}}_{g,n}$ is that it compactifies $\mathrm{M}_{g,n}$ without allowing the points to come together. When points on a smooth curve approach each other the curve sprouts off one or more components, each isomorphic to a projective line, and the points distribute themselves at smooth points on these new components (which is similar to the compactification $X[n]$ [**FM94**] of the space of n distinct points in a variety X).

We will focus on the case of genus zero. In this case, $\overline{\mathrm{M}}_{0,n}$ is a nonsingular variety. A point in $\overline{\mathrm{M}}_{0,n}$ corresponds to a curve which is a tree of projective lines meeting transversally, with n distinct, nonsingular, marked points; the stability condition is that each component must have at least three special points, which are either the marked points or the nodes where the component meets the other components.

Let X be a smooth projective variety, and let β be an element in $A_1(X)$. Let $\mathrm{M}_{0,n}(X,\beta)$ be the set of isomorphism classes of pointed maps $(C, p_1, \ldots, p_n, \mu)$ where C is a projective nonsingular curve of genus zero, the markings $p_1, \ldots, p_n$ are distinct points of C, and μ is a morphism from C to X satisfying $\mu_*([C]) = \beta$.

There is a compactification $\mathrm{M}_{0,n}(X,\beta) \subseteq \overline{\mathrm{M}}_{0,n}(X,\beta)$, whose points correspond to stable maps $(C, p_1, \ldots, p_n, \mu)$ where C a projective, connected, nodal curve of genus zero, the markings $p_1, \ldots, p_n$ are distinct nonsingular points of C, and μ is a morphism from C such that $\mu_*([C]) = \beta$. $(C, p_1, \ldots, p_n, \mu)$ is a stable map if the following condition holds for every irreducible component $E \subseteq C$:

If $E \cong \mathbb{P}^1$ and E is mapped to a point by μ, then E must contain at least three special points (either marked points or points where E meets the other components of C).

The simplest example is $\overline{\mathrm{M}}_{0,0}(\mathbb{P}^{r-1}, 1)$, which is the Grassmannian $\mathbf{G}_{2,r}$. If $n \geq 1$, $\overline{\mathrm{M}}_{0,n}(\mathbb{P}^{r-1}, 1)$ is a locally trivial fibration over $\mathbf{G}_{2,r}$ with the configuration space $\mathbb{P}^1[n]$ of [**FM94**] as the fiber [**FP97**].

Remark 6.5 ($\mathrm{M}_{0,n}$ and Chow Quotients). It is a result from [**Kap93**] that

$$\mathbf{G}_{2,n}(\mathbb{C}) /\!/ T^{n-1} = \overline{\mathrm{M}}_{0,n}(\mathbb{C})$$

where T^{n-1} is the maximal torus from $\mathbb{P}^{n-1}$ operating on the complex Grassmannian $\mathbf{G}_{2,n}(\mathbb{C})$. This construction was further investigated in [**GM07**], where it was shown that $\overline{\mathrm{M}}_{0,n}(\mathbb{C})$ equals the tropical compactification of $\mathrm{M}_{0,n}$, i.e. the compactification of $\mathbf{G}_{2,n}^{\circ}(\mathbb{C})/T^{n-1}$ in the toric variety defined by the fan trop $\mathbf{G}_{2,n}^{\circ}/T^{n-1} = \mathbf{G}_{2,n}^{\circ}(\mathbb{T})/L$ where L is the lineality space of the tropical Grassmannian.

Independent of this, the tropical $\mathrm{M}_{0,n}(\mathbb{T})$ had been defined in [**Mik06a**] as the set of all tropical curves of genus zero with n distinct marked points modulo homeomorphisms and was realized as a balanced polyhedral complex.

In [**GKM09**] the space $\mathrm{M}_{0,n}(\mathbb{T})$ was realized as a quotient of the tropical Grassmannian. It was noted in [**SS04**] that this quotient is homeomorphic to the space of phylogenetic trees constructed in [**LJBV01**].

In [**Mik06a**] Mikhalkin also introduced a compactification $\overline{\mathrm{M}}_{0,n}(\mathbb{T})$, where every edge length was allowed to become infinite. This is equal to the tropical compactification of $\mathrm{M}_{0,n}(\mathbb{T})$, where every ray ρ has an infinite end point in the torus orbit $O(\rho)$.

This means that tropical and classical $\mathrm{M}_{0,n}$ are linked by tropicalization:

Theorem 6.6.
$$\overline{M}_{0,n}(\mathbb{T}) = \text{trop}\,\overline{M}_{0,n}(\mathbb{K}).$$

It is a key feature of $M_{0,n}(\mathbb{T})$ that it is a hübsch fan, the underlying set $|M_{0,n}(\mathbb{T})|$ has a unique coarsest fan structure.

The spaces $M_{0,n}^{\text{lab}}(\mathbb{R}^r, d)$ have an r-dimensional lineality space, hence they do not have a canonical tropical compactification.

We know that the tropical Grassmannian $\mathbf{G}_{2,D+n}(\mathbb{T})$ is a compactification of the fan $\mathbf{G}_{2,D+n}^{\circ}(\mathbb{T})$. While some quotients of torus orbits correspond to limits of tropical curves, others do not corresponds to curves of the right degree, or not to any well-defined point set at all, since the map $\pi_{n,D}$ does not extend to a map $\mathbb{TP}(D+n) \to \mathbb{TP}(r)$.

A natural compactification in this context is the Chow quotient
$$\mathbf{G}_{2,D+n}(\mathbb{T}) /\!\!/ \ker \pi_{n,D} =: \overline{M}_{0,n}^{\text{lab}}(\mathbb{TP}(r), D).$$

1. Rational Tropical Curves of Degree One with n Marked Points

In this section we will investigate the space $\overline{M}_{0,n}^{\text{lab}}(\mathbb{TP}^{r-1}, 1)$. We begin by reviewing key properties of $\mathbf{G}_{2,k}^{\circ}(\mathbb{T})$ and $\mathbf{G}_{2,k}(\mathbb{T})$.

A generic point in $\mathbf{G}_{2,k}^{\circ}(\mathbb{T})$ represents a tree with k leaves and $k-1$ edges. Points in lower-dimensional cells correspond to trees with k leaves and fewer than $k-1$ edges.

A generic point in a matroid stratum $\mathcal{R}(M)$ represents a tree with A leaves and $A-1$ edges where A is the number of atoms in the matroid M.

This is a key difference between the compactification $\mathbf{G}_{2,k}(\mathbb{T})$ of $\mathbf{G}_{2,k}^{\circ}(\mathbb{T})$ and the compactification $\overline{M}_{0,n}(\mathbb{T})$ of $M_{0,n}(\mathbb{T})$: A generic point in $M_{0,n}$ represents a tree with n leaves and $n-1$ edges, but generic points in $\overline{M}_{0,n} \setminus M_{0,n}$ represent trees with n leaves and $n-1$ edges, some of which have infinite length.

Now let us look at points in $M_{0,n}^{\text{lab}}(\mathbb{R}^{r-1}, 1) = \mathbf{G}_{2,r+n}^{\circ}(\mathbb{T})/H_n$. It corresponds to a tree with two sorts of leaves:

(1) r leaves that represent edges in $\mathbb{R}^{r-1}$,
(2) n leaves that represent marked points on the tree.

This corresponds to two surjective maps:

- ft : $M_{0,n}^{\text{lab}}(\mathbb{R}^{r-1}, 1) \to \mathbf{G}_{2,r}^{\circ}$ forgets all information about the marked points and only keeps information about the tropical line in $\mathbb{R}^{r-1}$, i.e. the tree with r leaves above.
- pr : $M_{0,n}^{\text{lab}}(\mathbb{R}^{r-1}, 1) \to M_{0,n}$ forgets all information about the tropical line in $\mathbb{R}^{r-1}$ and only keeps information about the relative position of the marked points towards each other, i.e. the tree with n leaves above.

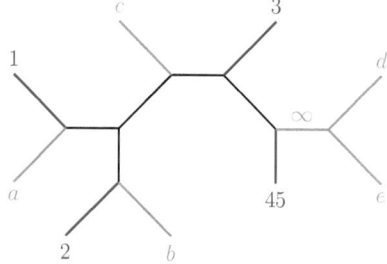

(a) A tree in the boundary of $\overline{\mathrm{M}}_{0,5}^{\mathrm{lab}}(\mathbb{TP}^4, 1)$.

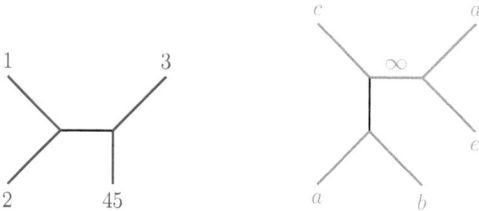

(b) A tree in the boundary of $\mathbf{G}_{2,5}(\mathbb{T})$. (c) A tree in the boundary of $\overline{\mathrm{M}}_{0,5}(\mathbb{T})$.

FIGURE 12. A point in the boundary of $\overline{\mathrm{M}}_{0,n}^{\mathrm{lab}}(\mathbb{TP}^{5-1})$ and its images under ft and pr in $\mathbf{G}_{2,5}$ and $\overline{\mathrm{M}}_{0,5}$. Rays in $\mathbb{R}^5/\mathbb{R}$ are labeled with $\{1,\ldots,5\}$ and the marked points are labeled with $\{a,\ldots,e\}$.

We will see that these maps extend to the compactifications, that is ft extends to a map $\overline{\mathrm{M}}_{0,n}^{\mathrm{lab}}(\mathbb{TP}^{r-1}, 1) \to \mathbf{G}_{2,r}(\mathbb{T})$ and pr extends to a map $\overline{\mathrm{M}}_{0,n}^{\mathrm{lab}}(\mathbb{TP}^{r-1}, 1) \to \overline{\mathrm{M}}_{0,n}(\mathbb{T})$.

Generic points in the boundary of $\overline{\mathrm{M}}_{0,n}^{\mathrm{lab}}(\mathbb{TP}^{r-1}, 1)$ will then correspond to trees with $A+n$ leaves where A is the number of atoms in a rank 2 matroid on r elements. The tree has $A + n - 1$ edges, some of which may have infinite length (see Figure 12. However, this is only possible for those edges of the tree which are contracted by the map from $\mathbb{R}^{r+n-1} \to \mathbb{R}^{r-1}$.

In order to show that $\overline{\mathrm{M}}_{0,n}^{\mathrm{lab}}(\mathbb{TP}^{r-1}, 1)$ is a parameter space of rational tropical curves with n marked points we need to describe the orbits of the toric variety $\mathbb{TP}\binom{r+n}{2}/\!\!/ H_n$.

We already know the matroid stratification of $\mathbf{G}_{2,r+n}(\mathbb{T})$. Every matroid stratum is the intersection of $\mathbf{G}_{2,r+n}$ with a torus orbit of $\mathbb{P}\binom{r+n}{2}$. The matroid strata are intersections with torus orbits of the ambient projective space. The Chow quotient $\mathbf{G}_{2,r+n}/\!\!/ H_n$ embeds into $\mathbb{P}\binom{r+n}{2}/\!\!/ H_n$. We know that the quotient of the central stratum $\mathbf{G}_{2,r+n}^{\circ}/H_n$ embeds into the Chow quotient $\mathbf{G}_{2,r+n}/\!\!/ H_n$. We will see that there are other strata whose topological quotient embeds into the Chow quotient.

If we want to determine these orbits, we have to describe the polytope P_n generated by the columns of a matrix A whose rows span the kernel of $\pi_{n,1}$.

Let us choose coordinates such that the kernel of $\pi_{n,1}$ is generated by the first n unit vectors of $\mathbb{R}^{r+n}/\mathbb{R}$. A unit vector e_i from $\mathbb{R}^{r+n}$ acts on $\mathbb{TP}\binom{\#D+n}{2}$ like the vector $\sum_{j\neq i} e_{ij}$.

Note that the choice of another basis leads to an affinely isomorphic polytope.

Let A be the $n \times \binom{r+n}{2}$ matrix defined via $a_{k,ij} = \begin{cases} 1, & k \in \{i,j\} \\ 0, & k \notin \{i,j\} \end{cases}$

This means the row space of A generates the kernel of $\pi_{n,1}$. We can describe the columns of A as follows:

- The vector zero appears $\binom{r}{2}$ times, for each combination $n < i < j$ of indices.
- The vector e_i appears r times, for each combination i,j with $i \leq r < j$ of indices.
- The vector $e_i + e_j$ appears exactly once for the combination i,j.

We see that all columns of A are vertices of conv A, so $\mathcal{R}(M)/H_n$ embeds into $\mathbf{G}_{2,r+n}$ whenever conv A_M = conv A.

The second hypersimplex Δ^2_{n+1} is isomorphic to the polytope with vertices e_i and $e_i + e_j$. The convex hull of all vectors e_i is a facet of the second hypersimplex in this coordinates. This leads us to the following description of the polytope P_n obtained as the convex hull of the columns A:

Lemma 6.7. P_n is an n-dimensional convex polytope with normalized volume $2^n - n$ and set of vertices $\{0\} \cup \{e_i\} \cup \{e_i + e_j\}$.

PROOF. The second hypersimplex Δ^2_{n+1} is n-dimensional and has normalized volume $2^n - n - 1$ [Stu95, Theorem 9.4]. The polytope P_n is the union of the second hypersimplex and the unit simplex with vertices $\{0\} \cup \{e_i\}$. □

Hence we see that an element Z of the Chow quotient $\mathbf{G}_{2,r+n}/\!/ H_n$ is a sum of torus orbits $Z = \sum \overline{H_n x_i}$ such that $\dim \overline{H_n x_i} = \dim \overline{H_n e_0} = n$ and $\sum \deg \overline{H_n x_i} = \deg \overline{H_n e_0} = 2^n - n$.

That means if there is a matroid M such that $\dim \overline{H_n e_M} = n$ and $\deg \overline{H_n e_M} = \deg \overline{H_n e_{U(n)}}$ then the topological quotient $\mathcal{R}(M)/H_n$ embeds into the Chow quotient $\mathbf{G}_{2,r+n}/\!/ H_n$ (where $U(n)$ is the uniform rank n matroid on n elements).

Example 6.8. For $n = 2$ the polytope P_2 is just the unit square (which has normalized volume $2 = 2^2 - 2$ since it can be decomposed into two lattice triangles). For $n = 3$ we get the 3-dimensional polytope from Figure 13 on the following page.

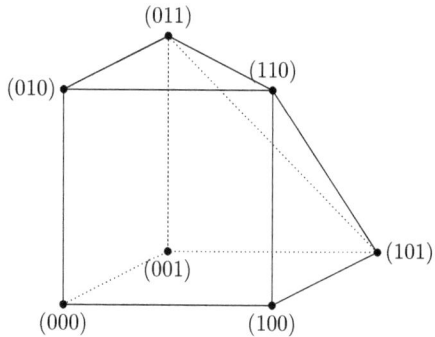

FIGURE 13. The polytope of H_3 is a subpolytope of the unit cube containing the second hypersimplex Δ_4^2.

The Chow quotient $\mathbb{P}\binom{r+n}{2}/\!/H_n$ corresponds to all subdivisions of P_n whereas the quotient $G_{2,r+n}/\!/H_n$ only depends on all matroid subdivisions of P_n. We can describe all matroid subdivisions of P_n

Lemma 6.9. *Let S be a matroid subdivision of P_n. Then S contains a polytope which contains the unit simplex.*

PROOF. S must contain a marked polytope (Q, M) which has a full-dimensional intersection with the unit simplex. This is only possible if Q contains the origin. Let i be a marked point. Then there must be a basis ij contained in M, for otherwise Q would lie in the hyperplane $x_i = 0$. By the basis exchange axiom M must contain a basis ki where k is not a marked point. Hence the unit vector e_i is contained in Q. Thus Q contains the unit simplex. □

Corollary 6.10. *A matroid subdivision of P_n determines a matroid subdivision of Δ_{n+1}^2 (and this is a matroid subdivision in the usual sense).*

The matroid subdivisions of the second hypersimplex describe the compactification $\overline{M}_{0,n+1}$ of $M_{0,n+1}$ and are known explicitly [**Kap93**, Theorem 1.3.6], we also know the geometry of $\overline{M}_{0,n+1}(\mathbb{T})$ explicitly. Note that the combinatorics of P_n with n marked points correspond to subdivisions of Δ_{n+1}^2, as we need to keep track of the position of the n marked points relative to each other and relative to the embedded tropical line.

Finally, let us have a look at the relation between $\overline{M}_{0,n}^{\text{lab}}(\mathbb{TP}^r, 1)$ and $\overline{M}_{0,n}^{\text{lab}}(\mathbb{TP}^{r-1}, 1)$. They are both defined by the polytope P_n, however in the former case the vertices occur with multiplicities 1, $r+1$ and $\binom{r+1}{2}$ while in the latter they are 1, r and $\binom{r}{2}$ respectively. In fact, the marked polytope $P_n = P_n(r)$ of $\overline{M}_{0,n}^{\text{lab}}(\mathbb{TP}^{r-1}, 1)$ is a matroid subdivision of the marked polytope $P_n(r+1)$ of $\overline{M}_{0,n}^{\text{lab}}(\mathbb{TP}^r, 1)$ and the corresponding matroid has $r+1$ as a loop. This implies that every

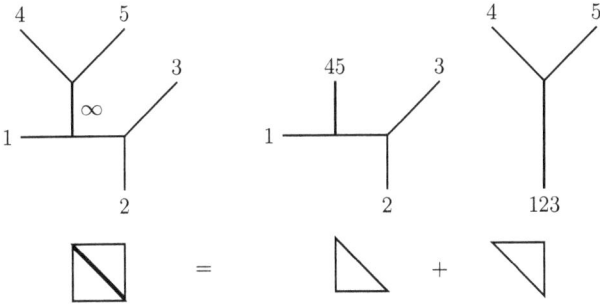

FIGURE 14. A tree with an infinite edge is the sum of two trees coming from a matroid subdivision.

subdivision of $P_n(r)$ is also a subdivision of $P_n(r+1)$, hence $\overline{\mathrm{M}}_{0,n}^{\mathrm{lab}}(\mathbb{TP}^{r-1},1)$ lies in the boundary of $\overline{\mathrm{M}}_{0,n}^{\mathrm{lab}}(\mathbb{TP}^{r},1)$. The same relation holds between $\mathbf{G}_{2,r}(\mathbb{T})$ and $\mathbf{G}_{2,r+1}(\mathbb{T})$, where $\mathbf{G}_{2,r}(\mathbb{T})$ consists of all strata $\mathcal{R}(M)$ of $\mathbf{G}_{2,r+1}(\mathbb{T})$ such that $r+1$ is a loop of M. This means that the subset of all tropical lines (with or without marked points) that lie in a torus invariant subvariety of $\mathbb{P}^r$ are the intersection of $\overline{\mathrm{M}}_{0,n}^{\mathrm{lab}}(\mathbb{TP}^{r},1)$ with a torus invariant subvariety of $\mathbb{P}\binom{r+n}{2}/\!\!/ H_n$ (or $\mathbf{G}_{2,r}(\mathbb{T})$ and $\mathbb{P}\binom{r}{2}$, respectively).

2. Rational Tropical Curves of Higher Degree Without Marked Points

Let us now look $\overline{\mathrm{M}}_{0,0}^{\mathrm{lab}}(\mathbb{TP}^{r-1},d)$ for $d \geq 2$. The situation is similar to the previous case with $d=1$ and $n>1$ in that boundary points in $\overline{\mathrm{M}}_{0,0}^{\mathrm{lab}}(\mathbb{TP}^{r-1},d)$ correspond to trees with fewer than rd leaves that may have contracted edges of an infinite length. In this case, however, such a contracted edge leads to a splitting of the curve into several components of lower degree.

Now, let us look at the polytopes involved in the quotient $\mathbb{TP}\binom{|D|}{2}/\!\!/ \ker \pi_D$. We begin with the case $d=2$, which has slightly different combinatorics then the general case $d \geq 3$.

The vectors e_i and e_{r+i} of $\mathbb{R}^{2r}$ get mapped to the same image for $i=1,\ldots,r$. Hence the kernel of $\pi_{0,2}$ is generated by all vectors $e_i - e_{i+r}$ for $i=1,\ldots,r$. A unit vector e_i from $\mathbb{R}^{2r}$ acts on $\mathbb{TP}\binom{\#D}{2}$ like the vector $\sum_{j \neq i} e_{ij}$.

Let A be the $r \times \binom{2r}{2}$ matrix defined via $a_{k,ij} = \begin{cases} 1, & k \in \{i,j\}, k+r \notin \{i,j\} \\ -1, & k+r \in \{i,j\}, k \notin \{i,j\} \\ 0, & \text{otherwise} \end{cases}$

This means the row space of A corresponds to the action of the kernel of $\pi_{0,2}$. We can describe the columns of A in $\mathbb{Z}^r$ as follows:

- The vector zero appears r times, for each combination $i, i+r$ of indices with $i \leq r$.
- The vector $e_i + e_j$ appears exactly once for the combination i, j with $i, j \leq r$.

- The vector $e_i - e_j$ appears exactly once for the combination $i, j+r$.
- The vector $-e_i - e_j$ appears exactly once for the combinations $i+r, j+r$.

The origin is the only interior lattice point of the convex hull $P^2 = \mathrm{conv}\{a_{ij}\}$ of the columns of A.

Let us now look at the general case of a labeled tropical degree D of s vectors in $\mathbb{Z}^{r-1}$: There appears to be no canonical choice for a basis of the kernel of the projection $\pi_{0,D}$. Let us assume we choose a $\binom{s}{2} \times$ matrix B whose columns generate the $\ker \pi_{0,D}$. The polytope P^D describing the quotient $\mathbf{G}_{2,s} /\!/ \ker \pi_{0,D}$ is then the convex hull of the rows of this matrix. It is of dimension $s-r$. We will now describe a symmetric generating system for the case of tropical degree d. It is given by the $rd \times rd$-matrix B with entries $b_{a,b,c,d} = \begin{cases} d-1, & \text{if } a=c \text{ and } b=d \\ -1, & \text{if } b=d, \text{ but } a \neq c \\ 0, & \text{otherwise} \end{cases}$

The matrix B describes the action $\ker \pi_{0,D}$ on $\mathbb{TP}(r)$. The action on $\mathbb{TP}(dr)$ is given via an $rd \times \binom{rd}{2}$-matrix A with entries $a_{a,b,a',b',c,d} = b_{a,b,c,d} + b_{a',b',c,d}$. We define the polytope P^d as the convex hull of the columns of A. Note that $P^d \subseteq \mathbb{R}^{rd}$ lies in an affine subspace of codimension r. We can easily derive a few facts from this description:

- No column is equal to the zero vector, though the origin is an interior point of P^d.
- Each column has at most $2r$ non-zero entries, the coordinates can be sorted into r blocks such that all coordinates from each block sum to zero.
- No column appears twice and no column is a convex combination of other columns.

Example 6.11. For $d=2$ and $r=3$ we get a three-dimensional polytope, the cuboctahedron (see Figure 15 on the next page). The combinatorics of the Chow quotient $\mathbb{P}\binom{2 \cdot 3}{2} /\!/ H^2$ depend on the fiber polytope $\Sigma(\Delta_{15}, P^2)$ which is an 11-dimensional polytope with 173232 vertices. The number of vertices (i.e. the number of regular triangulations of P^2) has been determined using the program TOPCOM [**Ram02**]. The torus H^2 for curves in $\mathbb{P}^2$ is three-dimensional just like the torus H_3 used for the space $\overline{\mathrm{M}}_{0,3}^{\mathrm{lab}}(\mathbb{TP}^2, 1)$. However, the fiber polytope $\Sigma(\Delta_{15}, P_3)$ only has 1296 vertices.

3. Forgetful Maps

In this section we will explore the relations between the compact spaces $\mathbb{TP}(r)$, $\overline{\mathrm{M}}_{0,n}(\mathbb{T})$ and $\overline{\mathrm{M}}_{0,n}^{\mathrm{lab}}(\mathbb{TP}(r), d)$.

The tropical parameter spaces $\mathrm{M}_{0,n}(\mathbb{T})$ and the classical moduli spaces $\overline{\mathrm{M}}_{0,n}(\mathbf{C})$ come equipped with forgetful maps

$$\mathrm{ft} : \mathrm{M}_{0,n+1} \to \mathrm{M}_{0,n}.$$

This forgetful map extends to the compactification $\overline{\mathrm{M}}_{0,n}(\mathbb{T})$.

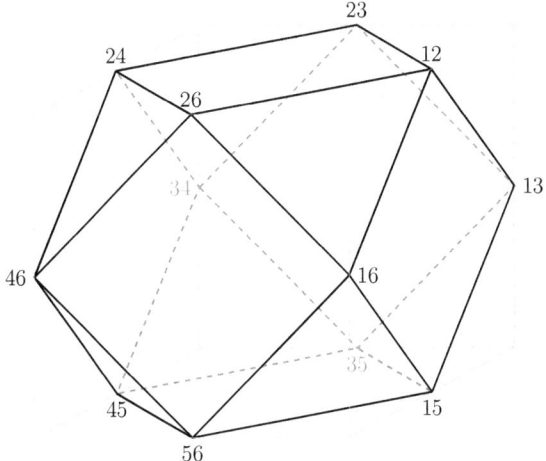

FIGURE 15. The polytope P^2 is the cuboctahedron. The six sides of the ambient cube $[-1,1]^3$ are labeled with numbers from 1 to 6 and every vertex of the cuboctahedron lies on the intersection of two sides of the ambient cube. The three pairs of opposing sides 14, 25, 36 correspond to the origin.

Theorem 6.12. *Let $r \geq 2$ and $n \geq 2$. Then there is a commutative diagram*

$$\begin{array}{ccc} \overline{M}^{\text{lab}}_{0,n+1}(\mathbb{TP}^{r-1},d) & \xrightarrow{\text{ft}} & \overline{M}^{\text{lab}}_{0,n}(\mathbb{TP}^{r-1},d) \\ \downarrow{\text{pr}} & & \downarrow{\text{pr}} \\ \overline{M}_{0,n+1}(\mathbb{T}) & \xrightarrow{\text{ft}} & \overline{M}_{0,n}(\mathbb{T}) \end{array}$$

where all horizontal maps are induced from projections $\mathbf{G}^\circ_{2,n+1} \to \mathbf{G}^\circ_{2,n}$ and all vertical maps from projections $\mathbf{G}^\circ_{2,r+n} \to \mathbf{G}^\circ_{2,n}$. The forgetful map

$$\text{ft}: \overline{M}^{\text{lab}}_{0,n+1}(\mathbb{TP}^{r-1},d) \to \overline{M}^{\text{lab}}_{0,n}(\mathbb{TP}^{r-1},d)$$

exists for all $n \geq 0$.

PROOF. We know there is a forgetful map ft : $\overline{M}_{0,n+1}(\mathbb{C}) \to \overline{M}_{0,n}(\mathbb{C})$, and hence a map $\overline{M}_{0,n+1}(\mathbb{K}) \to \overline{M}_{0,n}(\mathbb{K})$, and by tropicalization we get the desired map $\overline{M}_{0,n+1}(\mathbb{T}) \to \overline{M}_{0,n}(\mathbb{T})$. A direct proof that Chow quotients admit this map can be found in [**Kap93**, Theorem 1.6.6].

The forgetful map ft : $\mathbf{G}^\circ_{2,dr+n+1}/H^d_{n+1} \to \mathbf{G}^\circ_{2,dr+n}/H_n$ is defined via forgetting all Plücker coordinates of $\mathbf{G}^\circ_{2,dr+n+1}$ that use coordinate $n+1$. We will show that this map extends to the Chow quotients $\mathbf{G}_{2,dr+n+1}/\!/H_{n+1} \to \mathbf{G}_{2,dr+n}/\!/H_n$. The idea behind this proof is very intuitive: the forgetful map between varieties corresponds to the deletion of elements from matroids.

Let F_{dr+n+1} be the fan of $\mathbb{P}(dr+n+1)$ To show that ft defines a toric morphism, by Theorem 5.10 and Proposition 5.27 we need to show that the image of every cone of F_{dr+n+1}/H^d_{n+1} (determined by a matroid subdivision of P_{n+1} by $\Delta_{\binom{dr+n+1}{2}}$) gets mapped to a subset of cone of F_{dr+n}/H^d_n (induced by a subdivision of P_n by $\Delta_{\binom{dr+n}{2}}$).

Let S be a matroid subdivision of P^d_{n+1} and let $M_1, \ldots, M_k$ be the corresponding matroids. We consider the collection S' of all deletions $M'_i = M_i \setminus \{n+1\}$ that are of rank two. We want to show that this defines a matroid subdivision of P^d_n.

Let us identify $\mathbb{R}^n$ with $\mathbb{R}^n \times \{0\} \subseteq \mathbb{R}^{n+1}$. The vertices of P^d_{n+1} are the vertices of P^d_n plus the unit vector e_{n+1} and all sums $e_i + e_{n+1}$ for $i = 1, \ldots, n$. That means the collection of matroids $M_i \setminus \{n+1\}$ (some might be equal) naturally defines a subdivision of P_n when considered as a face of P_{n+1} (it is the intersection of P_{n+1} with the hyperplane $x_{n+1} = 0$).

The boundary fiber polytope of subdivisions of the boundary is a Minkowski summand of the fiber polytope [**BS92**, Prop. 3.1], hence we get a well-defined toric morphism

$$\text{ft}: \mathbf{G}_{2,dr+n+1} /\!\!/ H_{n+1} \to \mathbf{G}_{2,dr+n} /\!\!/ H_n$$

extending the map $\text{ft}: \mathbf{G}^\circ_{2,dr+n+1}/H_{n+1} \to \mathbf{G}^\circ_{2,dr+n}/H_n$.

Similarly, we need to show that the linear map $\text{pr}: \mathbf{G}^\circ_{2,dr+n}/H_n \to \mathbf{G}^\circ_{2,n}/H_n$ is a map of fans.

Again, the second hypersimplex Δ^2_n is a face of P^d_n, it is cut out by the equation $\sum_{i=dr+1}^{dr+n} x_i = 2$. Let σ be a cone of $\mathbb{P}\binom{dr+n}{2} /\!\!/ H_n$ corresponding to a matroid subdivision S. Let $M_1, \ldots M_k$ be the matroids of S. The restriction M'_i of M_i is a matroid on $[rd+n] \setminus [dr]$. We define the collection S' to contain all M'_i that are of rank two. This is the matroid subdivision of Δ^2_n induced by S. Therefore pr extends to the Chow quotients.

The commutativity follows from the fact that the maps $\text{ft} \circ \text{pr}$ and $\text{pr} \circ \text{ft}$ are identical on $\mathbf{G}^\circ_{2,dr+n+1}/H^d_{n+1}$. $\square$

Lemma 6.13. *Let $r \geq 2$ and $d \geq 1$. Then*

$$\mathbb{P}(dr) /\!\!/ H^d = \mathbb{P}(r).$$

PROOF. The linear map $\pi_d: (\mathbb{R}^r)^d \to \mathbb{R}^r$ is defined via $e_{i,j} \mapsto e_i$ for $i = 1, \ldots, r$, $j = 1, \ldots, d$.

Let B be the $(d-1)r \times r$-matrix with columns given by $b_{(i,j)} = e_{1,i} - e_{j,i}$. H^d is the kernel of this map. An orthogonal complement to the kernel is given by the image of $\mathbb{R}^r \to (\mathbb{R}^r)^d$, $e_i \mapsto \sum_j e_{i,j}$.

The $r \times dr$-matrix A with entries $a_{i,(k,l)=\delta_{ik}}$ describes a linear map $\mathbb{R}^r \to (\mathbb{R}^r)^d$ whose image is an orthogonal complement of the kernel of π_d. This means A is a Gale dual of B and the multi-intersections of the cones formed from the columns of A describe the secondary fan of the

80

polytope conv(B) (see [**LJBS90**, Section 4]). But the columns of A are just the unit vectors so the secondary fan is the fan of $\mathbb{P}(r)$. □

Another feature of the classical $\overline{\mathrm{M}}_{0,n+1}(\mathbb{CP}^{r-1}, d)$ and of $\mathrm{M}_{0,n}(\mathbb{R}^{r-1}, d)$ are evaluation maps that send a curve to the coordinates of one of its marked points.

Theorem 6.14. *Let $r \geq 3$ and $1 \leq i \leq n$. Then there is an evaluation map*

$$\mathrm{ev}_i : \overline{\mathrm{M}}_{0,n}^{\mathrm{lab}}(\mathbb{TP}^{r-1}, d) \to \mathbb{TP}^{r-1}$$

extending the map $\mathrm{ev}_i : \mathrm{M}_{0,n}(\mathbb{R}^{r-1}, d) \to \mathbb{R}^{r-1}$.

PROOF. We can assume $n = 1$ after applying forgetful maps. We will prove the theorem as follows

(1) $\mathbf{G}_{2,D+1} /\!/ H_1^D = (\mathbf{G}_{2,D+1} /\!/ H_1) /\!/ H^D$.
(2) $\mathbf{G}_{2,D+1} /\!/ H_1 = F\mathbf{G}_{2,D} \subseteq \mathbb{P}\binom{D}{2} \times \mathbb{P}(D)$. This follows from Lemma 5.23.
(3) There is a map $(\mathbb{P}\binom{D}{2} \times \mathbb{P}(D)) /\!/ H^D \to \mathbb{P}(D) /\!/ H^D$. This follows from Lemma 5.19.
(4) $\mathbb{P}(D) /\!/ H^D = \mathbb{P}(r)$. This follows from Lemma 6.13.

So all we need to show is $\mathbf{G}_{2,D+1} /\!/ H_1^D = (\mathbf{G}_{2,D+1} /\!/ H_1) /\!/ H^D$. We do know that there is a map $(\mathbf{G}_{2,D+1} /\!/ H_1) /\!/ H^D \to \mathbf{G}_{2,D+1} /\!/ H_1^D$ from Lemma 5.21. For our purposes we need a map in the opposite direction, though.

Let F be the fan of $\mathbb{P}\binom{D+1}{2}$ and let L^D, L_1 and L_1^D be the vector spaces corresponding to the tori H^D, H_1 and H_1^D respectively. Our first result is that every cone in F/L_1 is of the form σ/L_1 where σ is a cone of F, i.e. there are no multi-intersections. This because $\overline{H_1 e} \subseteq \mathbb{P}\binom{D+1}{2}$ is a subvariety of degree one, hence every effective cycle of the same degree as $\overline{H_1 e}$ consists of precisely one orbit closure.

Let us now check the the fan $(F/L_1)/L^D$. Every cone in there is a multi-intersection $\sigma = \bigcap \sigma_i / L^D$ where every σ_i is a cone of F/L_1. But that means $\sigma = \bigcap (\sigma_i/L_1)/L^D$ where every cone σ_i is a cone F. This is set-theoretically the same as the multi-intersection $\sigma = \bigcap \sigma_i/L_1^D$. And that means σ is a cone in F/L_1^D, i.e. there is a map $\mathbf{G}_{2,D+1} /\!/ H_1^D \to (\mathbf{G}_{2,D+1} /\!/ H_1) /\!/ H^D$. □

We actually believe that a key part of this proof holds true in more generality

Conjecture 6.15. $\mathbf{G}_{2,D+n} /\!/ H_n^D = (\mathbf{G}_{2,D+n} /\!/ H_n) /\!/ H^D = (\mathbf{G}_{2,D+n} /\!/ H^D) /\!/ H_n$.

The geometric meaning of this that the combinatorics of the marked points can be treated independently of the combinatorics of the combinatorics of the labeled degree and the embedding.

4. Interpretation of boundary points in $\overline{\mathrm{M}}_{0,n}^{\mathrm{lab}}(\mathbb{TP}^{r-1}, d)$

We do not fully understand the fan of Chow quotient $\overline{\mathrm{M}}_{0,n}^{\mathrm{lab}}(TP(r), d)$ nor do we know all matroid subdivisions of P_n^d.

We can, however, give examples of seven types of matroid subdivisions of P_n^d, we believe that every matroid subdivision can be achieved as a combination of those seven types.

We begin with a review of the fan $\mathrm{M}_{0,n}(\mathbb{T})$.

Definition 6.16. Let E be a finite set. An edge set S is a finite set of unordered pairs $\{A, B\}$ of subsets of E satisfying:

(1) Each $\{A, B\}$ describes a partition of E with $\#A, \#B \geq 2$.
(2) If $\{A, B\}$ and $\{C, D\}$ are in S then either $A \subseteq C$ or $C \subseteq A$ or $A \subseteq D$ or $D \subseteq A$.

Theorem 6.17. *The poset of cones of $\mathrm{M}_{0,k}(\mathbb{T})$ is isomorphic to the poset of edge sets of $[k]$ (ordered by inclusion). Let I, J be a partition of $[k]$ into elements of size at least two, then the ray generated by $r_{I,J} := \sum_{i,j \in I} -e_{ij}$ generates the cone corresponding to the edge set $\{\{I, J\}\}$. Note that $r_I = \sum_{i,j \in I} -e_{ij}$ is equivalent to $r_J = \sum_{i,j \in J} -e_{ij}$ modulo the lineality space of $\mathbf{G}_{2,k}^{\circ}$.*

PROOF. [SS04, Section 4]. □

This describes the cones of $\mathbf{G}_{2,k}^{\circ}(\mathbb{T})$, each cone is a cone of $\mathrm{M}_{0,k}$ plus the lineality space $T(k)$ of the tropical Grassmannian. Our ultimate goal is to find a fan structure for $\mathbf{G}_{2,rd+n}^{\circ}$ such that every cone τ of $\mathbf{G}_{2,rd+n}^{\circ}$ contains H_n^d in its lineality space and τ/H_n^d is contained in a cone of the fan of $\mathbb{P}\binom{dr+n}{2} /\!/ H_n^d$.

We are currently unable to provide such a fan structure. We will explore the necessary refinement of the lineality space $T(rd + n)$. We do this by choosing a factorization $T(rd + n) = T(r) \times H_n^d$ (H_n^d is by definition the kernel of the map $\pi_{n,d} : T(rd+n) \to T(r)$).

The factor $T(r)$ acts via translations of curves. We already know from translations of tropical lines that the fan on $T(r)$ will need to be a refinement of the fan of $\mathbb{P}(r)$. However, the fan structure that we need also depends on the degree d.

If v is a generic vector in $T(r)$ then the limit of translationg a curve C along v will be in some boundary orbit of $\mathbb{P}^{r-1}$ unless v is the negative of a ray generator of C, in which case the translation of that ray will stay in the main torus (this already happens with lines). The possible directions of rays of a curve depend on the degree. However, knowing just the rays does not determine the fan. We believe that the following construction yields a suitable fan.

Definition 6.18. Let D be a labeled tropical degree of curves in $\mathbb{TP}^{r-1}$. Then $F(D)$ is the normal fan of the Chow polytope of a generic curve of degree D.

Remark 6.19. The Chow polytope is the weight polytope of the Chow form. We did not define Chow forms for tropical complexes. One can either choose a generic non-Archimedean curve C whose tropicalization has degree D (this exists, as all tropical lines are realizable) and take the Chow polytope of C or use the results of [**Fin10**] which describes a direct construction for the normal fan of the Chow polytope from tropical complexes.

Definition 6.20. We define the fan $F(r,d)$ to be the common refinement of the fan of $\mathbb{P}(r)$ with all $F(D)$ such that D is labeled tropical degree of curves in $\mathbb{TP}^{r-1}$ which have projective degree d.

Unfortunately, we have not fully understood the combinatorics of the fan $F(r,d)$ either. The easiest non-trivial case is if $r = 3$ and $d = 1$, in which case $F(3,1)$ is the common refinement of the fan of $\mathbb{P}^2$ with its negative.

Let us investigate the limit points of the rays of $\overline{M}_{0,rd+n}$ and of vectors from $T(r)$:

(1) Let $-e_i$ be one of the rays of $\mathbb{P}(r)$. The limit point of this ray is a curve in the boundary divisor D_i of $\mathbb{P}(r)$. In the case of $d = 1$ this is the torus orbit corresponding to the matroid which has i as a loop (the corresponding subdivision uses all of P_n, but omits all vertices of the form ij). If $d \geq 2$, then there is a subdivision of P_n^d into d matroid polytopes, each polytope corresponds to the cycle which has one of the d copies of i as a loop. Each of these polytopes describes a curve of degree d in $\mathbb{TP}(r-1)$ with $d-1$ additional marked points – the other $d-1$ copies of i. An example of this can be seen in Figure 17 on page 85.

(2) Let v_i be the negative of one of the rays of a labeled degree D. The limit point of this ray is a curve which consists of ordinary lines in $\mathbb{R}^{r-1}$ (parallel to $\mathbb{R}v_i$) and possibly some components in the boundary. In the case of $d = 1$ and $n = 0$, this is the torus orbit of a matroid with only two atoms, i and $[r] \setminus \{i\}$. If we have $n > 0$ then this corresponds to a matroid subdivision of P_n into two polytopes, one consists of the matroid with two atoms, i and $[r+n] \setminus \{i\}$, the other has loops $[r] \setminus \{i\}$. This corresponds into the decomposition of P_n into a simplex and the second hypersimplex Δ_{n+1}^2, where the element i acts as an additional marked point (i.e. this component describes the position of the marked points with respect to the limit point of line $\mathbb{R}e_i$). See Figure 16 on the next page for an illustration of this. In the general case, we have a cycle $Z = \sum H_n^d x_i$ where the x_i are all limits of rays $\mathbb{R}_{\geq 0} v$ such that $\pi_{n,d}(v) = v_i$ and $\dim H_n^d x_i = \dim P_n^d$. If D is the degree of curves of tropical degree d, then this will be a sum of d irreducible components, each one describing a line in $\mathbb{R}^r/\mathbb{R}$.

(3) Let us look at the limit of the ray r_I in the case that I consists only of marked points. The corresponding edge gets contracted by $\pi_{n,D}$. The limit curve has an formal infinite length for this contracted edge. The subdivision consists of the matroid where all elements of I are parallel and the matroid where all elements not in I are

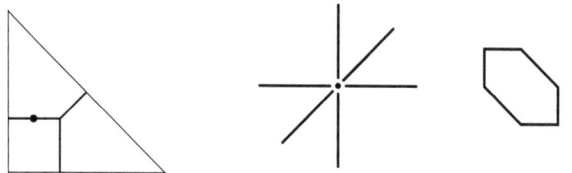

(a) A tropical curve of degree one with one marked point in $\mathbb{TP}^2$.

(b) The common refinement of the fan of $\mathbb{P}^2$ and its negative. It is the normal fan of a lattice hexagon.

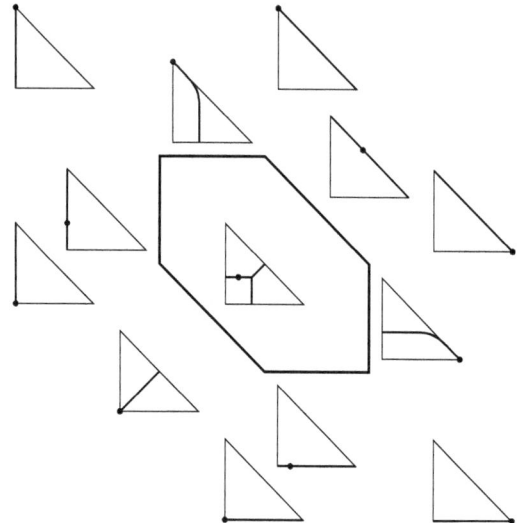

(c) Limit curves depend on the cone containing the vector used for translation.

FIGURE 16. Possible limits of a curve of degree one with one marked point under translation.

parallel. This is illustrated in Figure 18 on page 86. The corresponding cycle has two components, $\pi_{n,d}$ maps one component to a curve of degree d and the other gets contracted to a point (at the common position of the marked points indexed by I).

(4) Let us look at the limit of the ray r_I in the case that I consists only of marked points and precisely one direction i. This is a bounded edge of the curve who is parallel and adjacent to a ray of the curve. The bounded edge becomes infinitely long, and all marked points get moved to the limit point of the bounded edge. The corresponding subdivision consists of the matroid where all elements from I are parallel and the matroid where all elements not in I are parallel and i is a loop. As in the previous case,

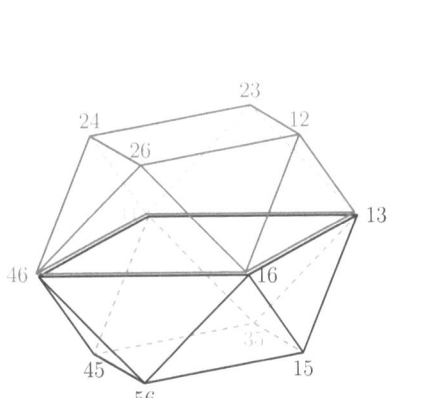

(a) A matroid subdivsion of P^2 into matroids with loops.

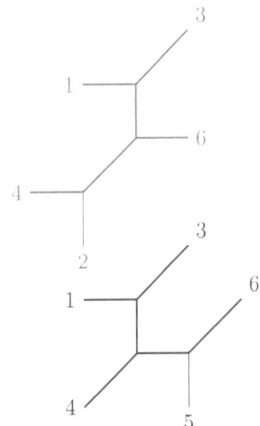

(b) A sum of cycles of $\mathbf{G}_{2,6}$ in the corresponding orbit of $\overline{M}_{0,0}^{\text{lab}}(\mathbb{P}^2, 2)$.

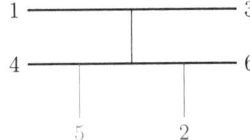

(c) The corresponding curve is of degree two in the boundary of $\mathbb{P}^2$ with two additional marked points.

FIGURE 17. A subdivision of P^2 corresponding to a curve in the boundary of $\mathbb{P}^2$.

the corresponding cycle has a component representing the curve and one component representing the marked point (which now sits outside the main torus).

(5) Let us look at the limit of the ray r_I in the case that I consists of marked points and two or more directions, but such that I without the marked points does not contain a tropical degree. In this case the corresponding edge of the curve gets stretched to become a ray, and all rays with directions in I vanish. This the image of a curve in the corresponding orbit of the tropical Grassmannian. If $d = 1$ then this a subdivision of P_n with just one polytope, which misses all vertices corresponding to pairs of elements from I. If $d > 1$ then this is a subdivision of P_n^d into the matroid polytope which is defined by the property that all elements from I are parallel and for every pair of two elements from I a matroid where all directions not from I but parallel to directions from I are loops. An example of this can be seen in Figure 19 on page 87.

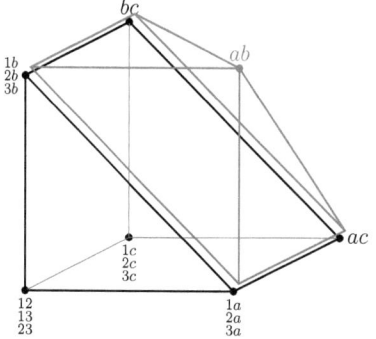
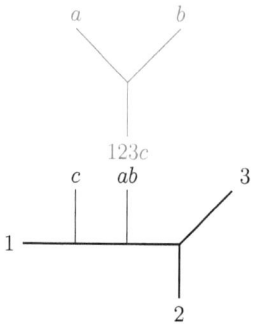

(a) A subdivision of P_3, the marked polytope for degree one curves in $\mathbb{P}^2$ with three marked points.

(b) A sum of cycles of $\mathbf{G}_{2,\{1,2,3,a,b,c\}}$ from the corresponding orbit of $\overline{\mathrm{M}}_{0,3}^{\text{lab}}(\mathbb{P}^2,1)$. Note that all edges of the top curve will be contracted.

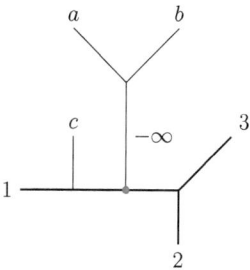

(c) The corresponding deformation of a tropical curve has a contracted edge of infinite length.

FIGURE 18. A polytope subdivision corresponding to a contracted edge of infinite length.

(6) Let us look at the limit of the ray r_I in the case that I consists marked points and directions such that I without the directions constitutes a tropical degree. In this case the corresponding edge gets contracted and the image curve has two components which are each of lower degree. The limit curve has a formal infinite length for this contracted edge, it consists of two curves of lower degree each with an additional marked point, the intersection point of the two components. The torus orbit corresponds to a matroid decomposition into two matroids, one where all elements from I are parallel and one where all other elements are parallel. This is illustrated in Figure 20 on page 88.

(7) Let us look at the limit of the ray r_I in the case that I consists of directions and marked points such that I without the marked points contains a tropical degree and additional

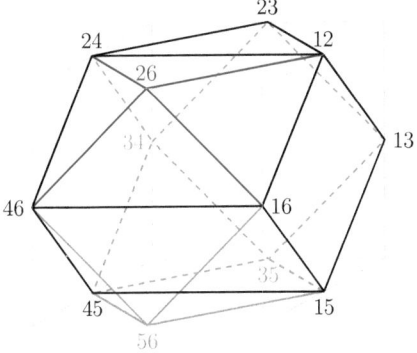

(a) A subdivision of P^2 into three matroid polytopes.

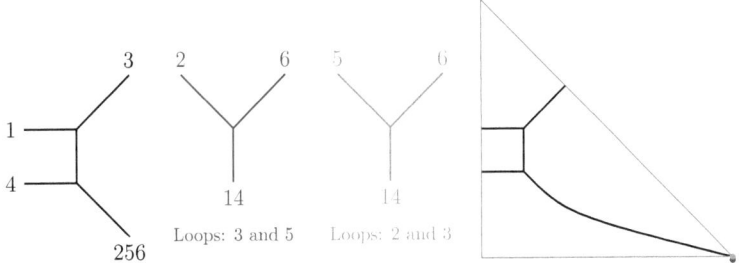

(b) The corresponding cycle has three components, two components correspond to lines in the boundary.

(c) One component of the cycle represents a curve, the other two get contracted to a point.

FIGURE 19. A matroid decomposition corresponding to a non-generic curve of projective degree two.

directions J. In this case the limit curve has two components of lower degree, one in $\mathbb{R}^{r-1}$ with a ray in direction $v = \sum_{j \in J} v_j$ and one in a boundary of $\mathbb{TP}^r$ (as if it where translated along v). Both have an additional marked point which corresponds to the intersection of the two components. The matroid subdivision is the appropriate combination of the previous type and the first and second types.

5. Applications to Algebraic Geometry

We can construct complex (and non-Archimedean) analogues of the parameter spaces $\overline{\mathrm{M}}_{0,n}^{\mathrm{lab}}(\mathbb{TP}^r, d)$. Let us begin with $\overline{\mathrm{M}}_{0,0}^{\mathrm{lab}}(\mathbb{CP}^{r-1}, d)$. Let L be a a generic line in $\mathbb{CP}^{r-1}$. Let π_d be the toric morphism $(\mathbb{C}^\times)^{rd}/\mathbb{C}^\times \to (\mathbb{C}^\times)^r/\mathbb{C}^\times$ defined via $z = z_{i,j} \mapsto (\prod z_{1,j}, \ldots, \prod z_{r,j})$. Then $C = \overline{\pi_d(L)}$ is a smooth curve of degree d in $\mathbb{CP}^{r-1}$.

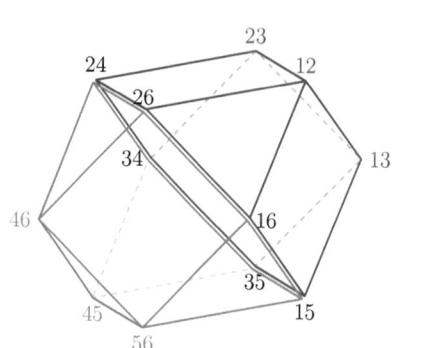
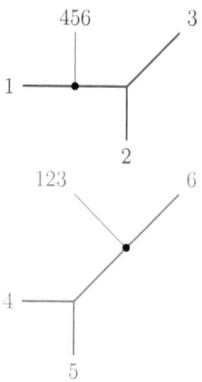

(a) A subdivision of the marked polytope P^2 of degree two curves in $\mathbb{P}^2$.

(b) A sum of cycles in the corresponding orbit. Note that the rays 123 and 456 will get contracted and each component of the cycle describes a curve of degree one.

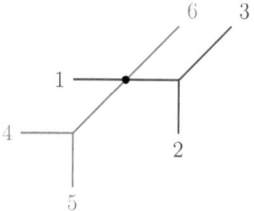

(c) The corresponding curve is connected and black!80!whiteucible, the irblack!80!whiteucible components meet in an additional marked point.

FIGURE 20. The polytope P^2 for labeled tropical curves without marked points in $\mathbb{P}^2$.

However, this construction provides more information than just the curve C. C will have d distinct intersection points with each boundary divisor D_i of $\mathbb{CP}^{r-1}$, each of these intersection points corresponds to a unique intersection point of L with one of the d boundary divisors of $\mathbb{TP}^{rd-1}$ that map to D_i under π.

88

Hence this construction produces a generic degree d curve C and a labeling of the intersection points of C with the boundary divisors of $\mathbb{P}^{r-1}$. This is the data contained in the quotient $\mathbf{G}^o_{2,rd}(\mathbf{C})/\ker \pi_d =: \mathrm{M}^{\mathrm{lab}}_{0,0}(\mathbf{CP}^{r-1}, d)$.

We can construct marked points by choosing a map $\pi_{n,d} : \mathbf{C}^{\times rd+n}/\mathbf{C}^\times \to \mathbf{C}^{\times r}/\mathbf{C}^\times$ which forgets the last n coordinates. We already investigated the corresponding Chow quotient and saw that this produces evaluation maps which define marked points.

Hence, for an algebraically closed field $\mathbb{K}$ we can define the projective variety

Definition 6.21. $\overline{\mathrm{M}}^{\mathrm{lab}}_{0,n}(\mathbb{K}\mathbb{P}^{r-1}, d) := \mathbf{G}_{2,rd+n}(\mathbb{K}) /\!\!/ \ker \pi_{n,d}$.

In the non-Archimedean case, the tropicalization of this space is $\overline{\mathrm{M}}^{\mathrm{lab}}_{0,n}(\mathbb{TP}^{r-1}, d)$. It is the Chow quotient whose combinatorics we have investigated for the tropical case. In the same way that we associated connected unions of tropical rational curves to every point in $\overline{\mathrm{M}}^{\mathrm{lab}}_{0,n}(\mathbb{TP}^r, d)$ we can associate connected nodal algebraic curves to every point in $\overline{\mathrm{M}}^{\mathrm{lab}}_{0,n}(\mathbb{K}\mathbb{P}^{r-1}, d)$ and hence turn $\overline{\mathrm{M}}^{\mathrm{lab}}_{0,n}(\mathbb{K}\mathbb{P}^{r-1}, d)$ into a parameter space of curves.

The next question is of course:
What is the relation between $\overline{\mathrm{M}}^{\mathrm{lab}}_{0,n}(\mathbf{CP}^{r-1}, d)$ and $\overline{\mathrm{M}}_{0,n}(\mathbf{CP}^r, d)$?

We do not have a proof for the general case, but low-dimensional examples strongly suggest:

Conjecture 6.22. $\overline{\mathrm{M}}^{\mathrm{lab}}_{0,n}(\mathbf{CP}^r, 1)$ is isomorphic to $\overline{\mathrm{M}}_{0,n}(\mathbf{CP}^r, 1)$.

The situation with degree $d > 1$ is more complicated.

As noted before, the variety $\mathrm{M}^{\mathrm{lab}}_{0,n}(\mathbf{CP}^{r-1}, d)$ is a parameter space of labeled degree d curves, that means that for a given generic curve through n distinct points there are rd different labelings. There is an corresponding action of the r-fold power of the symmetric group $(S_d)^r = S_d \times \cdots \times S_d$ on $\overline{\mathrm{M}}^{\mathrm{lab}}_{0,n}(\mathbf{CP}^r, d)$, one should therefore compare the spaces $\overline{\mathrm{M}}^{\mathrm{lab}}_{0,n}(\mathbf{CP}^{r-1}, d)/(S_d)^r$ and $\overline{\mathrm{M}}_{0,n}(\mathbf{CP}^{r-1}, d)$.

We already know (by the discussion in previous chapter) that these spaces are not equal: $\overline{\mathrm{M}}^{\mathrm{lab}}_{0,n}(\mathbf{CP}^{r-1}, d)/(S_d)^r$ depends on the toric structure of $\mathbf{CP}^{r-1}$, curves lying inside torus orbits are represented differently from curves lying in generic hyperplanes. Additionally, the subset of $\overline{\mathrm{M}}^{\mathrm{lab}}_{0,n}(\mathbf{CP}^{r-1}, d)$ consisting of curves that lie in a boundary divisor of $\mathbb{P}^{r-1}$ is not equal to (not even of the same dimension) $\overline{\mathrm{M}}^{\mathrm{lab}}_{0,n}(\mathbf{CP}^{r-2}, d)$.

Nonetheless, we know that we can interpret points in $\overline{\mathrm{M}}^{\mathrm{lab}}_{0,n}(\mathbf{CP}^{r-1}, d)/(S_d)^r$ as connected curves of degree d in $\mathbb{P}^{r-1}$, and boundary points correspond to connected projective nodal curves. The combinatorial structure of the marked points in the space $\overline{\mathrm{M}}^{\mathrm{lab}}_{0,n}(\mathbf{CP}^{r-1}, d)/(S_d)^r$ is the same as that of $\overline{\mathrm{M}}_{0,n}(\mathbf{CP}^{r-1}, d)$, the combinatorics are described via the maps to $\overline{\mathrm{M}}_{0,n}$.

Conjecture 6.23. There is a regular birational map from $\overline{\mathrm{M}}^{\mathrm{lab}}_{0,n}(\mathbf{CP}^{r-1}, d)/(S_d)^r$ to the moduli space $\overline{\mathrm{M}}_{0,n}(\mathbf{CP}^r, d)$.

Bibliography

[AK06] Federico Ardila and Caroline J. Klivans. The Bergman complex of a matroid and phylogenetic trees. *Journal of combinatorial Theory*, 96(1):38–49, 2006, arXiv:math/0311370v2.

[AN09] Daniele Alessandrini and Michele Nesci. On the tropicalization of the Hilbert scheme. 2009, arXiv:0912.0082v1.

[AR08] Lars Allermann and Johannes Rau. Tropical rational equivalence on $\mathbb{R}^r$, 2008, arXiv:0811.2860v2.

[AR09] Lars Allermann and Johannes Rau. First steps in tropical intersection theory. *Mathematische Zeitschrift*, 264(3):633–670, 2009, arXiv:0709.3705v3.

[Ber90] Vladimir G. Berkovich. *Spectral theory and analytic geometry over non-archimedean fields*, volume 33 of *Mathematical Surveys and Monographs*. American Mathematical Society, Providence, RI, 1990.

[BJS+07] Tristram Bogart, Anders Jensen, David Speyer, Bernd Sturmfels, and Rekha Thomas. Computing tropical varieties. *Journal of Symbolic Computation*, 42(1):54 – 73, 2007, arXiv:math/0507563v1.

[BLS+99] Anders Björner, Michel Las Vergnas, Bernd Sturmfels, Neil White, and Günter M. Ziegler. *Oriented Matroids*, volume 46 of *Encyclopedia of Mathematics and its Applications*. Cambridge University Press, Cambridge, second edition, 1999.

[BS92] Louis J. Billera and Bernd Sturmfels. Fiber polytopes. *Annals of Mathematics*, 135(3):527–549, 1992, http://www.jstor.org/stable/2946575.

[BS94] Louis J. Billera and Bernd Sturmfels. Iterated fiber polytopes. *Mathematika*, (41):348–363, 1994.

[Cox95] David A. Cox. The homogeneous coordinate ring of a toric variety. *Journal of Algebraic Geometry*, 4:17 – 50, 1995, arXiv:alg-geom/9210008v2.

[Cox01] David A. Cox. Minicourse on toric varieties. Lecture notes from a course given at the University of Buenos Aires, 2001, http://www.cs.amherst.edu/~dac/lectures/toric.ps.

[Ewa96] Günter Ewald. *Combinatorial Convexity and Algebraic Geometry*. Springer-Verlag, New York, 1996.

[Fin10] Alex Fink. Tropical cycles and Chow polytopes. 2010, arXiv:1001.4784v1.

[FM94] William Fulton and Robert MacPherson. A compactification of configuration spaces. *Annals of Mathematics*, 139(1):183–225, 1994, http://www.jstor.org/stable/2946631.

[FP97] William Fulton and Rahul Pandharipande. Notes on stable maps and quantum cohomology. *Proceedings of Symposia in Pure Mathematics*, 62(2):45–96, 1997.

[FR10] Georges François and Johannes Rau. Tropical intersections of cycles in matroid fans. to appear, 2010.

[FS97] William Fulton and Bernd Sturmfels. Intersection theory on toric varieties. *Topology*, 36(2):335 – 353, 1997, arXiv:alg-geom/9403002v1.

[Ful93] William Fulton. *Introduction to Toric Varieties*, volume 131 of *Annals of Mathematical Studies*. Princeton University Press, Princeton, New Jersey, 1993.

[GGMS88] Israel M. Gelfand, R. Mark Goresky, Robert D. MacPherson, and Vera V. Serganova. Combinatorial geometries, convex polyhedra and Schubert cells. *Advances in Mathematics*, 63:301–316, 1988, doi:10.1016/0001-8708(87)90059-4.

[GKM09] Andreas Gathmann, Michael Kerber, and Hannah Markwig. Tropical fans and the moduli spaces of tropical curves. *Compositio Mathematica*, 145(1):173 – 195, 2009, arXiv:0708.2268v1.
[GKZ94] Israel M. Gelfand, Mikhail M. Kapranov, and Andrei V. Zelevinsky. *Discriminants, Resultants and Multidimensional Determinants*. Birkhäuser, Boston, 1994.
[GM07] Angela Gibney and Diane Maclagan. Equations for Chow and Hilbert quotients. 2007, arXiv:0707.1801.
[GM08] Andreas Gathmann and Hannah Markwig. Kontsevich's formula and the WDVV equations in tropical geometry. *Advances in Mathematics*, 217:537, 2008, arXiv:math/0509628.
[Har92] Joe Harris. *Algebraic Geometry: a first course*. Springer-Verlag, New York, 1992.
[HKMP06] Megumi Harada, Yael Karshon, Mikiya Masuda, and Taras Panov, editors. *Toric Topology: International Conference May 28–June 3, 2006, Osaka City University, Osaka, Japan*. American Mathematical Society, 2006.
[Hu05] Yi Hu. Topological aspects of Chow quotients. *Journal of Differential Geometry*, 69(3):399–440, 2005, arXiv:math.AG/0308027.
[Kap93] Mikhail M. Kapranov. Chow quotients of Grassmannians I. I. M. Gelfand Seminar 16, Adv. Soviet Math., 29–110, 1993, arXiv:alg-geom/9210002v1.
[Kat09a] Eric Katz. Tropical intersection theory from toric varieties, 2009, arXiv:0907.2488v1.
[Kat09b] Eric Katz. A tropical toolkit. *Expositiones Mathematicae*, 27(1):1-36, 2009, arXiv:math/0610878v3.
[Kol96] János Kollár. *Rational curves on Algebraic Varieties*, volume 32 of *Ergebnisse der Mathematik und ihrer Grenzgebiete*. Springer, New York, 1996.
[KSZ91] Mikhail M. Kapranov, Bernd Sturmfels, and Andrei V. Zelevinsky. Quotients of toric varieties. *Mathematische Annalen*, 290(1):643–655, 1991, doi:10.1007/BF01459264.
[LJBS90] Paul Filliman Louis J. Billera and Bernd Sturmfels. Constructions and complexity of secondary polytopes. *Advances in Mathematics*, 83(2):155–179, 1990, doi:10.1016/0001-8708(90)90077-Z.
[LJBV01] Susan P. Holmes Louis J. Billera and Karen Vogtmann. Geometry of the space of phylogenetic trees. *Advances in Applied Mathematics*, 27(4):733–767, 2001, doi:10.1006/aama.2001.0759.
[LQ09] Mark Luxton and Zhenua Qu. On tropical compactifications. Preprint, 2009, arXiv:0902.2009.
[Mar07] Thomas Markwig. A field of generalized puiseux series for tropical geometry. 2007, arXiv:0705.2441,.
[MFK94] David Mumford, John Fogarty, and Frances Kirwan. *Geometric Invariant Theory*, volume 34 of *Ergebnisse der Mathematik und ihrer Grenzgebiete*. Springer, third edition edition, 1994.
[Mik06a] Grigory Mikhalkin. Moduli spaces of rational tropical curves. 2006.
[Mik06b] Grigory Mikhalkin. Tropical geometry and its applications. *International Congress of Mathematicians*, 2:827–852, 2006, arXiv:0601041.
[MMKZ92] Bernd Sturmfels Mikhail M. Kapranov and A. V. Zelevinsky. Chow polytopes and general resultants. *Duke Mathematical Journal*, 67(1):189–218, 1992.
[Oda78] Tadao Oda. *Lectures on Torus Embeddings and Applications*. Tata Institute of Fundamental Research, 1978.
[Oxl92] James G. Oxley. *Matroid Theory*. Oxford University Press, Oxford, 1992.
[Pay09a] Sam Payne. Analytification is the limit of all tropicalizations. *Math. Res. Lett.*, 16(3):543 – 556, 2009, arXiv:0805.1916v3.
[Pay09b] Sam Payne. Fibers of tropicalization. *Mathematische Zeitschrift*, 2:301, 2009, arXiv:0705.1732v2.
[Ram02] Jörg Rambau. Topcom: Triangulations of point configurations and oriented matroids. In Arjeh M. Cohen, Xiao-Shan Gao, and Nobuki Takayama, editors, *Mathematical Software–ICMS 2002*. World Scientific, 2002, http://www.zib.de/PaperWeb/abstracts/ZR-02-17. software at http://www.rambau.wm.uni-bayreuth.de/TOPCOM/.

[Rau09]   Johannes Rau. *Tropical intersection theory and gravitational descendants*. PhD thesis, TU Kaiserslautern, 2009, http://kluedo.ub.uni-kl.de/volltexte/2009/2370/pdf/Published.pdf.
[RGST05] Jürgen Richter-Gebert, Bernd Sturmfels, and Thorsten Theobald. First steps in tropical geometry. In G. L. Litvinov and V. P. Maslov, editors, *Idempotent Mathematics and Mathematical Physics*, volume 377, chapter Contemporary Mathematics, pages 289–317. American Mathematical Society, 2005, arXiv:arXiv:math.AG/0306366.
[SP05]   Bernd Sturmfels and Lior Pachter, editors. *Algebraic Statistics for Computational Biology*. Cambridge University Press, 2005.
[Spe05]   David Speyer. *Tropical Geometry*. PhD thesis, UC Berkeley, 2005, arXiv:math.CO/0311370.
[Spe08]   David Speyer. Tropical linear spaces. *SIAM Journal on Discrete Mathematics*, 22(4):1527–1558, 2008, arXiv:math/0410455.
[SS04]   David Speyer and Bernd Sturmfels. The tropical grassmannian. *Advances in Geometry*, 4(3):389 – 411, 2004, arXiv:math/0304218.
[Stu93]   Bernd Sturmfels. *Algorithms in Invariant Theory*. Springer-Verlag, Wien, 1993.
[Stu95]   Bernd Sturmfels. *Gröbner Bases and Convex Polytopes*. American Mathematical Society, 1995.
[Stu02]   Bernd Sturmfels. *Solving Systems of Polynomial Equations*. American Mathematical Society, 2002.
[Tev07]   Jenia Tevelev. Compactifications of subvarieties of tori. *American Journal of Mathematics*, 129(4):1087–1104, 2007, arXiv:math/0412329v3.
[Zie95]   Günter M. Ziegler. *Lectures on Polytopes*. Springer-Verlag, Berlin, 1995.

I want morebooks!

Buy your books fast and straightforward online - at one of world's fastest growing online book stores! Environmentally sound due to Print-on-Demand technologies.

Buy your books online at
www.morebooks.shop

Kaufen Sie Ihre Bücher schnell und unkompliziert online – auf einer der am schnellsten wachsenden Buchhandelsplattformen weltweit! Dank Print-On-Demand umwelt- und ressourcenschonend produziert.

Bücher schneller online kaufen
www.morebooks.shop

KS OmniScriptum Publishing
Brivibas gatve 197
LV-1039 Riga, Latvia
Telefax:+371 686 204 55

info@omniscriptum.com
www.omniscriptum.com

Printed by Books on Demand GmbH, Norderstedt / Germany